From Warfare to Welfare

PUBLISHING FOR THE WORLD
125 Years
THE JOHNS HOPKINS UNIVERSITY PRESS

From Warfare to Welfare

Defense Intellectuals and Urban Problems in
Cold War America

Jennifer S. Light

The Johns Hopkins University Press
Baltimore and London

This book has been brought to publication with the generous
assistance of the Pribram Fund.

The Johns Hopkins University Press
2715 North Charles Street
Baltimore, Maryland 21218-4363
www.press.jhu.edu

Library of Congress Cataloging-in-Publication Data

Light, Jennifer S., 1971–
  From warfare to welfare : defense intellectuals and urban
problems in Cold War America / Jennifer S. Light
    p. cm.
Includes bibliographical references and index.
  ISBN 0-8018-7422-X (alk. paper)
1. Urban policy—United States—History—20th century.
2. Federal-city relations—United States—History—20th century.
3. Technology and state—United States—History—20th
century. 4. National security—United States—History—20th
century. 5. United States—History—1945– I. Title.
  HT123.L45 2003
  307.76′0973′0904—dc21   2003006229

A catalog record for this book is available from the British
Library.

Contents

Acknowledgments

This book had its origins in a summer I spent working at the RAND Corporation. I joined a team working on new methods for defense science and technology planning, designing a WWW-based tool for collaborative public policy decision making. The plan was to use this technology to lead military officials through a decision-making environment and then to model the consequences of their choices—and alternatives—in the context of several different wartime scenarios.

Part of my job was to prepare a literature review of findings on the historical role of information technology in collaborative decision making. The conclusions I drew were entirely unexpected. Rather than finding that decades of investments had produced definitive knowledge about how such tools improved decision-making processes, I concluded that in all but a few cases, the results of such technologies were murky. My mentors at RAND urged me to publish the findings, but I declined. Perhaps I was missing something, I thought; surely, so many resources would not be devoted time and again to trying to improve on a category of innovations whose benefits repeatedly remained unproven?

It is precisely that question—How and why are resources allocated time and again to support the adoption of technical and technological tools whose benefits remain unproven?—that motivated me to write this book. I focus on how several decades of American city planners and managers came to rely on innovations first developed to fight the cold war. The RAND Corporation plays a starring role in this history, and I am indebted to Robert Lempert and James Bonomo for offering me the opportunity to see firsthand the inner workings of that fascinating institution.

Equally essential to the start-up of this project was the Graham Foundation for Advanced Studies in the Fine Arts. Support from the Graham Foundation is acknowledged in remarkably many of my favorite books on the history and theory of architecture and urban planning. I am honored to be able to thank viii *Acknowledgments*

the foundation here for generously sponsoring much of my research travel and all of the visual material costs of this project.

Historians whose focus is the recent past occasionally have the good fortune to meet some of the men and women who have made history, and several of the participants in this historical story generously gave their time for extended conversations. Harry Finger patiently answered many of my questions during a chance encounter at a New Year's Eve party, and in a later, more structured conversation, helped me to understand more precisely how both Democrats and Republicans saw hope for the future of America's cities in the products of the nation's defense and space programs. M. C. Branch invited me into his home to share his recollections and let me read his as-of-yet unpublished autobiographical musings on the history and future of planning. Leland Johnson, Laurence Lynn, and Henry Rowen each offered answers to questions I could not figure out based on documentary records alone and pointed me toward additional sources to incorporate into the book. Speaking with these men in person and on the telephone confirmed what I had hoped in undertaking this project—that in many cases, efforts to transfer defense and aerospace innovations to address the nation's urban problems were undertaken with good intentions.

Interdisciplinary research thrives in a community that makes collegiality across departments a priority, and Northwestern University provided an ideal home in which to write this book. Colleagues across the campus, including Jonathan Caverley, James Ettema, Susan Herbst, Eric Klinenberg, James Schwoch, and Marc Ventresca all volunteered to read a draft version of this book, and the final result reflects their extensive comments and criticisms. John Hudson, whose career at Northwestern's Geography Department spanned the era of defense- and space-agency sponsorship for the kinds of research described in this book, helped to compensate for decades of department records having been lost in a flood.

Northwestern University also provided extensive institutional support for this project. Colleagues in the Department of Communication Studies made all efforts to arrange my teaching schedule to facilitate productive research. At the School of Communication, Dean Barbara O'Keefe extended the honor of Ameritech Research Professor for a year, giving me the time and funding to complete the book. At the Institute for Policy Research, Fay Cook warmly welcomed me to join the institute's interdisciplinary community of scholars as a faculty fellow. Funding from the University Research Grants Committee provided additional support for research travel.

The insights of colleagues at several other institutions also significantly shaped the pages of this book. Michael Dudley sent his then-unpublished research on the defensive dispersal movement. Stephen Graham offered comments on an article-length version of my arguments. Robert Bruegmann, John Cloud, Michael Gordin, Richard John, Nicholas King, Ronald Kline, Bill Leslie, Richard Light, and John Durham Peters each offered feedback on an earlier version of the manuscript. Peter Buck read a draft manuscript as he has read nearly everything I have ever written, and our conversations continue to be a highlight of my return visits to Cambridge.

While this project was not based on my dissertation, so many of the skills I learned in graduate school, and people I met in those years, laid the intellectual foundations on which this project could be built. Sherry Turkle, Everett Mendelsohn, and Peter Buck were scholarly models par excellence. Fellow graduate students, including Eileen Anderson-Fye, Michael Gordin, Kristen Haring, Edward Jones-Imhotep, David Kaiser, Nicholas King, and Deborah Weinstein, continue to be sounding boards for ideas.

Historical research is impossible without the work of librarians and archivists, and several stand out for special mention. At the Los Angeles City Archives, Jay Jones met me as planned on the morning of September 11, 2001, and kept the archives open when most city buildings had closed. At the RAND Corporation, Vivian Arterberry, Ann Horne, and Roberta Shanman helped me to locate the information I was looking for in record time. At the Municipal Archives of the City of New York, Kenneth Cobb and Leonora Gaitlin offered astute advice about files in the John Lindsay collection. At Northwestern University Library, Victoria Zabohlsky and the team of librarians in interlibrary loan tracked down obscure conference proceedings and limited-circulation reports. In doing so, they significantly reduced the amount of research travel needed to complete this project.

At the Johns Hopkins University Press, this project was ably assisted by Robert J. Brugger, who urged me to write history. Our conversations convinced me that historical inquiry has an important, if undervalued, role to play in current affairs, and I hope this book will find its way to both historians and city administrators. Melody Herr shepherded the manuscript through the review and publication process, making every step a pleasure. Dennis Marshall offered superb advice about polishing and tightening my arguments. An anonymous referee provided several helpful comments, both substantive and organizational, that strengthened the final product.

Finally I must mention the personal relationships that sustain any professional exercise. My father, Richard, the other Professor Light, taught me the value of simplicity in academic writing. My mother, Patricia Light, reminded me that if I took breaks for fun I would finish the book sooner. My sister, Sarah Light, provided moral support and additional incentives for research trips to New York City. My grandmother Dede, whose entire American lifetime has been spent in New York City, shared her own memories of the period. Jonathan Caverley gave me eight years of seeing the U.S. military far more up close and in person than I ever expected to. This book is dedicated with love and appreciation to him.

From Warfare to Welfare

Introduction

At the 1966 meeting of the National League of Cities, the league's president, Detroit Mayor Jerome Cavanaugh, called attention to a troubling contradiction of the era in his opening speech. "Our readiness to jump into wars when they are outside the three-mile limit seems much greater than our readiness to jump into wars inside our national boundaries," he observed. Federal spending was continuing to favor the defense and space agencies over domestic programs, and Cavanaugh cautioned this was a narrow and shortsighted view of how to provide for the nation's security. In his estimation, "an equal threat" to the security of the United States could be found at home. In a recent three-month period, Mayors had confronted thirty-eight urban rebellions. Yet federal appropriations still implied that "the guerilla warfare in the Mekong Delta" was sixteen times more important than "the guerilla warfare on our city streets."[1]

Cavanaugh accompanied his critique with a proposed solution. Earlier in the decade, he noted, the nation's political leaders had committed themselves to putting a man on the moon, had appropriated adequate funding, and by 1966 this "target" lay in sight. Observing that federal antipoverty efforts appeared to offer ideal defensive "weapons" to combat the nation's domestic troubles, Cavanaugh, together with the league's Resolution Committee, urged Congress to commit to the War on Poverty with equal resolve by creating a ten-year urban redevelopment fund to mobilize national support for the needs of cities.[2]

The next speaker on the program perceived Mayor Cavanaugh's remarks as public slaps in the face. James Webb, that year's president of the American Society for Public Administration, was also the top official at the National Aeronautics and Space Administration (NASA). "I didn't come here to challenge the priorities of the American city," Webb defended, "and I would like to say that neither I nor any other leader in the space program has ever suggested that it should have any priority over the needs of the American city."[3] This meeting was not the first where Webb had spoken about urban issues, but his address got off to an especially rocky start.

The exchange between Cavanaugh and Webb underscored a period of complicated interactions between U.S. city administrators and the nation's defense and aerospace communities. Beginning in the mid-1960s, the Vietnam War and the Apollo Program became symbolic targets for city planners and managers who argued that society's spending priorities were misplaced during an era of urban crises at home. However, at the same time that some were denouncing profligate defense and aerospace spending as detracting from more urgent matters, many urban administrators looked to the military-industrial complex for guidance. From Mayor Cavanaugh's call to mobilize a national effort to combat urban poverty on the model of the space program to New York City Mayor John Lindsay's efforts to import military management expertise from the RAND Corporation to streamline city operations, even those big-city mayors who publicly shamed the excessive resources committed to the military-industrial complex simultaneously saw in those investments potential opportunities to improve cities. Webb's invitation to speak to an audience of thirty-five hundred mayors and other city officials reflected the era's popular view that America's defense and aerospace communities possessed essential knowledge to be shared with city planners and managers facing crises on domestic soil. So, too, did the award that the National Academy of Public Administration later named for Webb, celebrating contributions to urban research and management.

In fact, collaborative relationships between America's military-industrial complex and its city planners and managers already had begun to take shape two decades earlier, in the years immediately following World War II. In a climate of concerns about reducing urban vulnerability to atomic attack, military strategists, urban planners, atomic scientists, social welfare advocates, and local government officials came together for a sustained conversation about improving the nation's physical and social infrastructure in the postwar period. The social networks these civil defense discussions created centered around fears of external threats lay the groundwork for a new type of collaboration in the decades that followed, collaborations that would refocus the attention of military and urban planners and managers toward new fears about internal threats to the nation's security: urban problems.

In the decade following World War II, another set of anxieties about the state of American cities moved to center stage alongside fears of urban vulnerability to atomic attack. Traffic, poverty, overpopulation, and crime appeared to be worsening, despite the nation's rising standards of living. Federal programs, most prominently urban renewal, tried to steer U.S. cities on a course toward prosperity. Yet early efforts fell far short of their goal. By the late 1950s, exasperated urban planners and managers were seeking new directions for urban reform.

Military planners and managers in industry and government suspected they might have something to offer their colleagues in city administrations. Investments in defense and aerospace research and development already had spawned a variety of innovations whose potential applications to supervising complex and large-scale systems seemed nearly limitless.[4] As it would happen, defense research institutions such as RAND and SDC and aerospace companies such as Lockheed and McDonnell recently had decided that the long-term survivability of their organizations depended on finding ways to transfer these innovations beyond military markets. City planning and management quickly emerged as targets of opportunity. The proposal, a more systematic and scientific approach to city administration, seemed to promise a remedy to the urban professions' grand public failure.

At the U.S. Housing and Home Finance Agency (later the Department of Housing and Urban Development) and in cities across the nation, administrators were captivated by the promise of more scientific planning and management tools. Yet lacking in-house familiarity with the tools, they required some assistance. Beginning in the early 1960s, experts from think tanks and aerospace companies found themselves recruited to serve as advisers to American city governments. From cybernetics to computer simulations to satellite reconnaissance, techniques and technologies originally developed for military users in the 1940s, 1950s, and 1960s thus became the focus of efforts to better plan and manage U.S. cities in the 1960s and 1970s.

At first, the transfer of defense and aerospace innovations to urban operations proceeded apart from any sense of a war on city problems. The partnership seemed made in heaven—a more scientifically sound approach to planning and management for cities and more contracts for the defense and aerospace community. These ongoing efforts at market expansion then unexpectedly received a booster shot and a new rationale from national political events: the escalation of urban crisis.

No official wars were fought in the continental United States during the twentieth century. Yet as Cavanaugh suggested in his remarks to the National League of Cities, the urban riots of the 1960s came perilously close. As a tradition of nonviolent protests gave way to more militant protests in the latter part of the decade, city leaders faced civil-rights demonstrations, urban riots, and conflicts over Vietnam. The "long, hot summers" from 1965 to 1968 saw more than three hundred episodes of civil disorder, resulting in two hundred deaths and the destruction of several thousand businesses. Threats of bombing and other acts of sabotage escalated, presenting mayors and law enforcement across the nation with situations increasingly out of their control. In several cases, quelling urban riots became a domestic job for military troops.[5]

Yet members of the armed forces and the National Guard keeping order in U.S. city streets during these disturbances provided merely a short-term link between the defense and aerospace communities and city governments during the cold war. Alongside such public maneuvers were less visible examples of how military strategy and national security expertise were called upon in a longer-term effort to bring order to America's increasingly racially divided urban cores. As former *Newsweek* correspondent Samuel Yette discussed in his book *The Choice* (1971), following urban riots the House UnAmerican Activities Committee pressured President Lyndon Johnson to declare martial law in U.S. cities. Johnson, instead of calling for a full-scale domestic deployment of military troops, recruited an army of "defense intellectuals"—civilian scientists and social scientists from top universities, think tanks, and aerospace companies.[6]

Defense intellectuals from institutions such as RAND and Lockheed, already seeking urban markets, found further opportunities for work in the violence that had engulfed American cities. By framing the urban crisis as a national security crisis, their task became "civil defense" of a new variety: maintaining domestic urban security by continuing to apply defense and aerospace innovations and ideas to city planning and management. Cities, federal agencies, think tanks, and foundations followed Johnson's lead, creating numerous fora to bring together defense intellectuals to analyze the causes of urban violence and disorder and to prevent them in the future. At the Kerner Commission, at a RAND Workshop on Urban Problems sponsored by the Ford Foundation, and at meetings on urban technology organized by the American Institute of Aeronautics and Astronautics, a shared vision of the escalating "urban crisis" as a national security crisis—the same turn of phrase Cavanaugh had used to call for an end to excessive defense and aerospace spending—helped to transform urban problems into strategic challenges to be met by defense intellectuals deploying techniques and technologies of command, control, communications, computers, intelligence, and reconnaissance.[7]

Cold War Cities and the Military-Industrial Complex

The appeal of cities as residential areas has waxed and waned throughout American history. In the three decades immediately following World War II, the middle classes deemed many U.S. cities decidedly undesirable places to live. Historians describing the period have devoted significant attention to the discovery and rediscovery of "urban problems," and public and private efforts to solve those problems, alongside the simultaneous burgeoning of a middle-class suburban landscape.[8]

To date, however, accounts of American urban history have overlooked how two of the era's defining features, the cold war and the growth of a military-industrial complex, intersected with the approaches that federal and local leaders chose to address the complex problems they identified in the postwar period. A few historians have documented how the military-industrial complex served as an economic engine for urban and suburban physical change. Studies of the development of suburban housing for returning GIs, the role of the National Interstate and Defense Highway Act in creating an interstate system, and the ways some municipal leaders seeking funding for local priorities profited from increases in defense spending, describe a physical reshaping of the landscape in line with military priorities—or at least how planning projects undertaken for other reasons in the cold war era could be made to line up with the rhetoric of national defense.[9]

Yet such accounts, which have offered insights into relationships between urban and suburban physical and social change, have remained silent about the rise of a new class of urban experts, men (and indeed they were almost exclusively men) whose personal experience working for military sponsors led them to identify connections between the challenges faced by military and urban planners and managers. In a climate of a perceived crisis in urban administration, and fearing that urban problems presented threats to domestic order, many American city planners and managers turned to these men, and to the nation's military-industrial complex, for advice and inspiration. The same individuals and institutions who rose to prominence developing strategies to protect the nation from atomic attack thus found several decades of work guiding domestic responses to urban problems. A central focus for their efforts was the application of defense and aerospace techniques and technologies to urban operations.[10] Innovations originally designed to combat America's foreign enemies overseas and at home became the weapons of choice in battles to solve urban problems and maintain security in the nation's cities.

Tracing the migrations of individual experts and the evolution of defense and aerospace institutions alongside their transfer of specific techniques and technologies to several cities, this book presents evidence to suggest that a new narrative, one in which the military-industrial-academic complex and technical and technological developments inside city administrations become central, deserves to assume its place alongside other themes in American urban history.[11] In this narrative, a different set of actors, the bevy of technological enthusiasts referred to as "defense intellectuals"—civilian scientists and social scientists who were employed by the defense establishment—play starring roles. Their positions as advisers to government long outlasted their efforts to transfer any specific innovation.

From Warfare to Welfare offers a retelling of American urban history. Its story about the adoption of innovations in local government, and the complexities of technology transfer, brings perspectives from the history of science and technology to the American urban context. Historians of science and technology have identified the cold war and the growth of a military-industrial-academic complex as defining features of twentieth-century U.S. science and engineering. Defense- and space-agency initiatives such as the Manhattan Project and the Apollo Program created "big science" endeavors to pool intellectual resources from across disciplines. Federal sponsorship offered financial incentives to reorient university researchers toward the study of topics for mutual gain. And institutions from think tanks to aerospace companies provided new fora for civilian researchers to apply their knowledge in the service of national defense.[12] Large-scale investments in defense and aerospace research and development thus brought to America's security establishment an arsenal of new techniques, technologies, and institutions.

Despite their origins in a culture marked by secrecy, many cold war–era developments in science and engineering remained exclusive accessories to the defense establishment only temporarily. From systems analysis to satellites to think tanks, these innovations soon were adopted and adapted for civilian applications in both public and private sectors. While the story has not yet been part of conventional accounts of American urban history, city planning and management were no exception.

The products of defense and aerospace research and development influenced a variety of operations in American cities, from transportation planning to crime control to emergency management. This book focuses on several of their more unexpected influences: in community development ventures, comprehensive planning efforts, and projects to facilitate communication among citizens. The book's title refers to its overarching theme: that during the cold war, strategies for urban problem solving were heavily influenced by, and in some cases directly derived from, military techniques and technologies originally used against America's foreign enemies. Experiments in the nation's cities adapted the expertise of defense professionals to face new enemies: urban chaos, blight, and unrest. Seven chapters take the reader from World War II to 1975, documenting three decades of collaborations among defense and aerospace experts and urban planners and managers. Their stories reveal how the rise of a military-industrial-academic complex offered these collaborators professional prestige, research funding, and hope for maintaining order in U.S. cities.

City planning has been and remains an activity of both public and private organizations—for example, municipal government, developers, chambers of commerce, and private design firms. This book focuses on activities inside city governments. Throughout, a focus on applications in New York City and Los Angeles details the effects of military innovations and expertise in specific urban settings. Despite geographic, administrative, and cultural differences undermining the precision of terms such as *urban* and *city*, both New York City and Los Angeles have come to represent for the second half of the twentieth century what Chicago was to the sociologists of the first half—each an archetypal American city that scholars have used to make claims about trends in the character of urban life across the nation.[13] Taking a fresh look at the familiar faces of these two much-studied metropolises, together with discussion of related developments in smaller cities (Pittsburgh and Dayton are two examples), illustrates how the defense and aerospace community shaped the intellectual history of city planning and management as academic disciplines, the organizational development of American cities, and the day-to-day practices of city administrators in surprising and important ways.

A critical finding of this book is that applications of military innovations and expertise to urban problems rarely served as sources of solutions. Defense and aerospace executives and engineers found new employment as consultants to cities and federal urban programs. Think tanks and aerospace companies found new civil systems contracts. University scholars found military sponsorship for urban research. City administrators, both Democrats and Republicans, found new approaches to management. Yet average city dwellers found few visible effects.[14] In city after city, for innovation after innovation, few experiments achieved their promised reforms. The lasting significance of this episode for U.S. urban history instead lies in its creation and maintenance of an urban "power elite" whose influence on the ways Americans conceptualize cities and their problems has persisted to the present day.[15]

In this retelling of American urban history, focusing on links between defense and aerospace innovations and urban life, the book also outlines an alternative history of what contemporary scholars characterize as "cybercities" —metropolitan areas where media spaces and physical spaces converge; where communications infrastructure is as important as gas, electricity, sewers, and water; where citizens, businesses, and government are linked into multiple communications networks.[16] Conventional accounts trace cybercities' origins to an emerging "culture of simulation" in the decades following World War II. This book suggests that the history of cybercities is better understood through an appreciation of simulations in the context of military war games at RAND and MIT than in the context of theories about postmodernity from Jean Baudrillard. Ironically, it was during a period of postwar antiurbanism, when military innovations were brought to bear on urban problems, that the early seeds of vibrant cybercities were sown.

The arguments of this book, while historical, should prove useful to contemporary urban planners and managers. By embedding the theme of defense and aerospace technology transfer in narratives of American urban development, *From Warfare to Welfare* offers insights to city administrators contemplating the adoption of military innovations old and new—computer simulations, global positioning systems, geographic information systems, and the internet. For while the defense intellectuals' actual proposals rarely worked out as they had hoped, their legacy still remains with us.

CHAPTER ONE

Planning for the Atomic Age

Creating a Community of Experts

In a presentation to the American Municipal Association in November 1945, University of Chicago sociologist Louis Wirth asked a question that was on many minds in the months following the destruction of Hiroshima and Nagasaki: "Does the atomic bomb doom the modern city?" Reflecting on the science of atomic destruction, much of it developed at his own institution, he suggested the answer to this question was no.[1] Wirth cited the size and growth pattern of the nation's cities and the administrative changes needed to disperse populations and industries away from downtowns vulnerable to attack. He concluded that such a massive reshaping of urban form could not reasonably be achieved in a short amount of time. Whether or not a bomb might "doom" the modern city in the event of an atomic attack, the existence of atomic weapons should not drive American city planning. According to Wirth, the realistic strategy for assuring urban security was world control of atomic weapons—not defensive city planning.

Wirth's resistance to dispersal is unsurprising, given that his life's work revolved around neighborhood-based studies of concentrated urban culture. Yet this opinion leader in so much urban research would cultivate less of a following in his conclusions about the ideal form for postwar cities. In the years immediately following World War II, a remarkable amount of expert attention began to focus on the question of what cities should look like in the atomic age. Defense experts, atomic scientists, urban planners, and public officials united around the idea of "defensive dispersal"—that deliberate dispersal of population and industries could reduce cities' vulnerability to attack.

That any scholar's resistance to defensive dispersal lacked broad appeal fits neatly with standard accounts of American urban development. Classic narratives of American urban and suburban history have long emphasized the theme of urban disintegration in the postwar period. In the decades from 1945 to 1975, myriad forces dispersed the nation's populations and industries. The federal urban renewal program gutted urban cores. The national highway program cut wide swaths through many cities. Private developers created suburbs such as Levittown for returning GIs. Suburban and regional malls decimated street life and commercial districts of downtowns. Middle-class whites fled cities for the safety of suburbs and private communities. Whether they have emphasized the primacy of public policies, private developers, specific building types, creditors, industry, or simply the desire of the middle classes, urban and suburban historians have placed the stories of sprawl and fear of the city at center stage in their accounts of American postwar urban development.[2]

The actual dispersal of America's twentieth-century urban physical landscape owes far more to the range of factors that scholars already have identified than to the defensive dispersal movement. However, as a force acting on the American urban professional landscape, the movement's impact was significant and lasting. While in 1945 Louis Wirth argued against a national dispersal program on account of the accompanying need for centralized governance and planning, many other urban professionals, from planners to politicians, viewed the proposal to align city planning with the nation's security needs as a great opportunity. As dispersal planning provided for the national defense, simultaneously it seemed to promise solutions to many city problems that urban leaders had identified in the prewar period—including traffic, congestion, and slums. Defense rationales for dispersal offered strategic rhetorical backing to bolster political support for comprehensive postwar planning, for increased federal aid to cities, for the continuing professionalization of planning and urban administration, and for increased funding to urban research.

It is this story of dispersal that served as a turning point in American urban history. For in these conversations about dispersal, a new approach to city planning and management, dedicating military expertise to the nation's urban needs, began to take shape. The military-industrial-academic complex that infiltrated so many aspects of life in the cold war would also guide American approaches to addressing urban problems.

From the walled cities of ancient Rome and the Renaissance to nine-teenth-century Paris, where Baron Georges Eugène Haussmann created boule-vards to facilitate the movement of troops through the city, concerns about

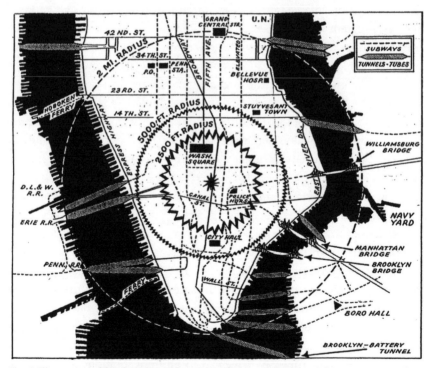

Fig. 1. Illustrations like this one of lower Manhattan accompanied numerous articles written by supporters of the defensive dispersal movement. This sketch depicts projected damage zones following an atomic blast, revealing how many of the city's escape routes would be unusable in the event of an attack. It appeared in a discussion of transportation planning for national defense in Edward Conway, "A-bomb over Manhattan," which appeared in *America Magazine* (July 22, 1950). Reprinted with permission of America Press, Inc. © 1950. All rights reserved.

Fig. 2. New York City Hall ca. 1943 displays civil defense emblems and a sign characterizing New York City as "Target No. 1." Reprinted with permission of the Municipal Archives, Department of Records and Information Services, City of New York.

national security have played a role in shaping the urban environment. In Europe, with its centuries of recorded military history and where cities from London to Berlin to Budapest have had to recover from bombing attacks, urban scholarship has framed cities past and present in conversation with war and military innovation. By contrast, studies of American urban development, planning, and administration have focused hardly at all on the ways that national security needs and military innovation have shaped the fate of specific American cities.[3]

Yet military needs and tools shaped the American urban experience in the postwar period, even if not as visibly as medieval fortification or redevelopment following air raids. Strategic bombing across Europe during World War I and the devastation in Japan during World War II offered evidence suggesting that cities and civilians were potential targets for a future nuclear strike. American civilians, like military troops, were warned to prepare for possible attack (figs. 1 and 2). A civil defense program with federal, state, and local dimensions brought these preparations for future war to the national consciousness. During the 1940s and 1950s, these civil defense initiatives offered important social settings for several professional groups—defense experts, atomic scientists, urban planners, and city managers—to come together in conversation about topics from highway planning to shelter design to future city form. Above all, it was discussions about dispersal planning that united their interests in physical and social planning. By 1953, city planner Coleman Woodbury called dispersal "the major issue in nonmilitary defense measures against the destruction of urban life and property by atomic attack."[4] These discussions, and the social networks they created, paved the way for technology-focused collaborations in the decades that followed.

Dispersal as Defense: Beginning the Conversation

America's atomic scientists were the first professional community to broach the topic of urban dispersal. Many of the same researchers whose innovations had helped to end World War II would warn of the consequences of their creations for dense urban concentrations and the need to protect civilians from outside threats. Even before atomic bombs were dropped on Hiroshima and Nagasaki, scientists working at the Met Lab (the Metallurgical Laboratory) at the University of Chicago initiated the discussion in the Franck Report, issued in July 1945 (a condensed version was subsequently published in December in the first issue of the *Bulletin of the Atomic Scientists,* a journal that would become a leading voice of arms control). In October 1945, the Association of Los Alamos Scientists issued a similar statement on atomic policy emphasizing "the vulnerability of our major industrial centers to atomic attack in the event of a future war."[5]

The question of how, and to what extent, to disperse concentrated urban areas became a major point of discussion for many of the nation's most distinguished scientists and science policy advisers with expertise in nuclear strategy. From Edward Teller (professor of physics at the University of Chicago and an architect of the bomb) to Ralph Lapp (a physicist who worked along with Teller on the Manhattan Project), key figures in the development of atomic weapons became interested in how urban policies during the postwar period might contribute to reducing civilian vulnerability to attack.[6] These scientists understood that the only defense against atomic weaponry was "absence from the locality in which it strikes."[7]

Views of this academic elite were fortified by support from high-level military officials and their colleagues in government and think tanks. Experts on strategic bombing theory echoed the atomic scientists and advised urban leaders to consider dispersal as their blueprint for civil defense. General Henry Arnold cautioned Americans in 1945 that the next attack might "be centralized on Michigan Boulevard, Biscayne Boulevard, Sunset Boulevard or on Main Streets in your home town." Three years later, Maj. Gen. Harold R. Bull told a group of assembled mayors that "no group of leaders could be more important to the essential planning and organization necessary to meet a threat of attack in the event of war, than the nation's chief Municipal executives." Lt. Gen. Leslie Groves, commanding officer of the Manhattan Project, wrote widely about the value of dispersal both for civilians and for industry. And military strategists at the U.S. Air Force think tank RAND, including a group calling itself the Dispersal Team, initiated several studies of the costs and benefits of different approaches to decentralizing the U.S. population.[8]

Captivated by warnings from atomic scientists and military strategists, the emerging profession of urban planners was quick to recognize that its experts' voices should become central to this conversation about future city planning. The American Society of Planning Officials (ASPO) had hosted conferences and issued official reports on defense topics as early as 1940. Yet like many other professional organizations, ASPO suspended meetings during the war. In the postwar period, planners and urban scholars from ASPO, as well as the American Institute of Planners, the National Planning Association, and the American Institute of Architects, created new committees and task forces on civil defense. The planning community began to speculate publicly about how its expertise might contribute to a more robust defense on the American home front.

Burnham Kelly, MIT professor of city planning, explained how the threats faced by urban populations had come full circle. Historically, cities had been constructed for "protection against warring neighbors," with the most salient features of urban design being "walls and moats."[9] Urban threats had evolved along with human civilization, he recounted, and police, fire, and hospital services had taken charge of civilian protection. Yet among the lessons of Pearl Harbor was a return to an older way of thinking about civilian protection—that once again, urban designs would have to place fear of military attack front and center. Who could be more useful to the civil defense discussion than city planners and managers whose careers had been devoted to studying the practicalities of urban design? The dispersal movement blossomed with their participation.

An astonishingly long list of city planners and urban scholars joined the dispersal conversation through research and testimonials about relationships between city design and national security. Many would find this topic ideally suited to securing research funding from military, government, and even private sponsors. Among them, Ansley Coale, a Princeton University graduate student, analyzed the matter in a report for the Social Science Research Council's Committee on the Social and Economic Aspects of Atomic Energy; University of Chicago sociologists and demographers Philip Hauser, Otis Dudley Duncan, and Beverly Duncan prepared a report on urban analysis for the U.S. Air Force Human Resources Research Institute; and together, three federal agencies—the National Security Resources Board (NSRB), the Department of Defense (DoD), and the Federal Civil Defense Administration (FCDA)—coordinated Project East River, a war-game study in which New York City, the nation's most concentrated urban center, was attacked by a bomb. (Such a concern was not unfounded; on Saturday July 28, 1945, a B-25 bomber had crashed into the Empire State Building.) Directed by Otto Nelson, a vice president at New York Life Insurance and a consultant on many city redevelopment projects, ten planners from the American Institute of Planners (AIP) were invited to collaborate as consultants, alongside Jerome Wiesner and other distinguished scientists. The planners included, among others, Coleman Woodbury, professor of regional planning at Harvard University; Tracy Augur, assistant to the director and planner in the Regional Studies Department of the Tennessee Valley Authority (TVA) and past president of the American Institute of Planners; William Wheaton, chair of regional planning at Harvard; Calvin Hamilton, a graduate student in regional planning at Harvard; Kelly, the professor of city planning at MIT and director of the Albert Farwell Bemis Foundation; and C. McKim Norton, a regional planner who served as president of the Regional Plan Association from 1940 to 1968. Each of these studies backed dispersal as a foundation for postwar planning. The question was how best to achieve it.[10]

Thus, at the same time that atomic scientists and military experts became ever more interested in questions of environmental design, city planners and managers became ever more interested in national security issues. Their intellectual exchange is evident in the professional literature of both communities in the 1940s and 1950s, where fears about the susceptibility of American cities to nuclear attack were the subject of articles advocating dispersal and urban planning to enhance national defense. During this period, the *Journal of the American Institute of Planners (JAIP)*, the voice of professional planning, and the *Bulletin of the Atomic Scientists* played host to the dispersal conversation (in 1951, the *Bulletin of the Atomic Scientists* devoted two issues to dispersal). Atomic scientists such as Lapp and planners such as Augur published across fields in both journals. The crossover was so significant that on several occasions a single article or position paper—for example, Augur's "The Dispersal of Cities as a Defense Measure" (1948)—appeared in both outlets. Similarly, in June 1953, the American Institute of Planners adopted a statement called "Defense Considerations in City Planning" at its annual meeting. In September, the statement was reprinted in the *Bulletin of the Atomic Scientists*. Discussions in the professional literature offer the most compelling evidence of emerging collaborations between these communities and the ways their disparate interests could be satisfied by uniting around the issue of postwar dispersal planning.[11]

In its early years, the significance of the dispersal movement lay in how it brought together previously distinct professional communities, mobilizing their resources to achieve a jointly defined goal: planning for the national defense. As the movement thrived, its significance for these professions also grew. City planners increasingly realized that defensive dispersal could contribute to far more than national security, or even to funding their research. By the early 1950s, a critical mass of planners understood how planning for urban dispersal offered an opportunity to advance other long-standing interests. A look at contributions to the dispersal conversation from several of the most prominent figures in the field offers insight into how urban professionals capitalized on concerns about urban security to advance arguments for their own prewar goals. These goals included some specific projects, such as reducing traffic and congestion, and slum clearance. They also included hopes for realizing utopian plans such as Ebenezer Howard's Garden Cities, a design for the English landscape developed in the late 1800s as a response to that nation's industrialization. As one anonymous planner put it, "We need not wait for a long range Russian bomber to teach us dramatically what street congestion, multiple parking, overbuilding, and lack of open express arteries can do to inspire fear, panic and unreasoning public fury."[12]

The Politics of Planning in the Atomic Age

City planning has a long history in the United States. Planning's position within local government, however, was largely a postwar innovation. Prior to that time, local chambers of commerce and prominent community business leaders largely directed local planning efforts. Planning commissions with varying degrees of political autonomy also guided the process. Few cities had more than one professional planner on staff. The American Institute of Planners had only 171 members in 1940, and a 1942 survey by the International City Managers' Association showed that half of U.S. cities spent no funds on city planning during the preceding year, including some of the nation's largest cities.[13]

Facilitated by the federal government's National Resources Planning Board and deepening relations between military experts and urban leaders in a wartime climate, during the 1940s planning became institutionalized in municipal government. As federal programs allocated money for civil defense and war production, they pushed local governments to administer these programs. For example, the Lanham Act funded more than four thousand public works projects between 1941 and 1946. As Thomas Hanchett has documented, mayors quickly discovered how local priorities could find funding if they could demonstrate relevance to the broad goal of strengthening national defense. By the late-1940s, surveys showed that city planning departments across the nation had expanded their staff and programs.

Historians such as Thomas Hanchett have stressed the importance of business interests in shaping the evolution of postwar planning. Yet the growing alliance between urban planning and management and national security strategy also would have lasting effects. If many cities did not have professional planners on staff until the postwar period, then professional practices in city agencies evolved in conversation with cold war civil defense concerns. Just as businesses and city mayors could find federal money for local public works projects using defense rationales—and indeed business publications such as *Fortune* and city mayors such as Milwaukee's Frank Zeidler were vocal dispersal proponents—so, too, planners such as Tracy Augur, Robert Moses, Burnham Kelly, and Catherine Bauer found that defensive dispersal offered additional political backing and a rhetoric of urgency to their prewar planning interests.[14] By suggesting to audiences how dispersal—properly implemented—simply extended some of the greatest planning theories of all time, city planners developed robust arguments that along with reducing a population's vulnerability to urban attack, their proposed city designs would improve the quality of urban life.

Augur was perhaps the most prolific city planner to contribute to the defensive dispersal discussion. Augur moved easily between urban and military professional communities, consulting for city planning departments across America (this included a prominent role in developing Oak Ridge, Tennessee), working at the NSRB, directing the Urban Targets Division of the Office of Defense Mobilization, and serving as president of the AIP and chair of its Committee on Defense Considerations in City Planning. Between 1946 and 1954, Augur delivered numerous speeches and wrote many articles declaring the need for rapid mobilization of resources to protect the U.S. population.[15]

Using the language of deterrence, Augur urged his audiences to consider city planning as a weapon. He compared urban development in other nations to the state of affairs in the United States, and cautioned that without the armament of sound city design, the United States was leaving itself susceptible to attack.[16] Augur reasoned that cities' centrality in the American economy looked much like Pearl Harbor as a concentration of naval power. The national network of cities was the mainstay of the country's current ability to function under wartime conditions. Yet if an urban structure vulnerable to attack were "likely to invite it," then urgent action to alter this structure would be necessary, whether the end result was (naming three popular dispersal plans of the day) "cluster cities," "ribbon cities," or "linear cities."[17]

In his arguments for dispersal, Augur called for a move to comprehensive military-style, top-down postwar planning from the national level, an idea many city planners had embraced during the prewar period. Augur appealed to practicality, noting that building to facilitate dispersal of urban populations would cost little more than traditional urban development, since building was always going on somewhere. By contrast, he argued, it would be far more costly to retrofit American urban centers, fixing built-in errors as an afterthought.

Augur's style of argument generally began from a focus on national security and then moved on to pair opportunity with danger. He took on those who argued that "the high levels of industrial production and organizational efficiency needed for modern war or modern peace could not be maintained if we were to break up such metropolitan concentrations as Detroit, Chicago, New York or Washington" by diagrammatic experiments to suggest otherwise.[18] Augur outlined many possible economic and social benefits that would follow from dispersal planning; for example, reduced congestion and blight; easier intra-urban travel; and a strong economy across the nation. Explaining that "before the atomic era" American cities had offered neither "good living" nor "protection from destruction, either by blight or bombs," he suggested that decentralization, always desirable, was even more so now.[19] Citing Howard's Garden Cities, Augur implied that such plans were not new but rather the brainchild of one of the greatest planners of all time. City planner Burnham Kelly (who participated in the military-sponsored Project Troy in addition to Project East River) and other dispersal proponents echoed Augur's claims, citing how the values allied with dispersal mobilization efforts were already central to urban planners' peacetime ways of thinking.[20]

Augur's invocation of Howard reveals how, for many planners, any call to disperse the American urban population "for the national defense" in fact was simply a new rationale fastened onto a much older argument. Long captivated by the idea of Garden Cities—the topic of his 1921 graduate thesis—Augur's career in the years prior to World War II already had been devoted to advancing Howard's ideals of master planning and dispersal in urban and regional design. For example, Augur had participated in regional planning efforts at the Regional Planning Association of America (along with such distinguished figures as Bauer and Clarence Stein) and worked at the TVA's early planning team.[21] What differed about the argument for dispersal during the postwar period was that it enabled federal agencies, national planning organizations, and individual cities to latch onto older planning theories in the name of defensive measures. In their enthusiasm for defense planning, they offered Augur and other dispersal supporters numerous public platforms.

Like Augur, Bauer, an influential figure in twentieth-century planning and housing, found the dispersal conversation offered ammunition to advance some of her older ideas. Like Augur, Bauer was an early member of the Regional Planning Association, also long committed to Garden City principles. Bauer served as vice president of the National Housing Conference, as a member of the Advisory Committee of the Division of Slum Clearance and Urban Redevelopment of the Housing and Home Finance Agency, and as a professor at Harvard University and the University of California, Berkeley. Jane Jacobs lists her name alongside Lewis Mumford and Clarence Stein as key disciples of Ebenezer Howard, suggesting that her level of commitment to decentralization was ideological and functioned apart from any sense of how well such plans actually worked.[22]

In keeping with the heritage of local planning in chambers of commerce, many planners tended to emphasize the economic benefits of dispersal in their public presentations—for example, holding that existing city congestion was causing "us losses even before the first bomb has been dropped," a perspective supported by a number of academic social scientists.[23] By contrast, Bauer's commentary focused in on the possibility of eradicating slums and improving life for the poorest urban dwellers. Her arguments aligned with a parallel tradition of planning as social reform dating to the Progressive Era. In her contribution to a redevelopment study edited by Coleman Woodbury, Bauer argued that, historically, crises such as epidemics and social revolution had served as catalysts for social welfare planning. She proposed that fear of nuclear attack become the crisis to set postwar planning in motion—that the nation's growing interest in defensive dispersal become the impetus for a comprehensive national planning program mobilized at the level of a military operation. Britain had done just this, she wrote, and she called for similar leadership from U.S. federal officials. Improving race relations was her first order of business. Second was increasing dispersal or "decentralization." In Bauer's mind, the two were closely linked. By superseding existing cities, a nation of garden cities would achieve civil defense goals and simultaneously solve a range of urban problems.

The crux of Bauer's argument was that because planning shapes environments for the long term, social welfare planning needs must be addressed alongside physical planning needs. Pointing to an overemphasis on postwar suburban planning for the middle classes, she urged planners and federal officials to refocus their attention on the fate of the inner city, where low-income citizens faced housing shortages and slum conditions.[24] Bauer politely told her readers that, despite rhetorical and even some public policy commitments to dispersal, the consequences of a continuing preference for high-rise public housing in American cities left poor and minority citizens as bomb targets, while the rest of the population was dispersed. Kelly made similar observations about the focus of the nation's urban renewal program on redevelopment. In his words, "public housing and even urban redevelopment, to say nothing of the tremendous projects of private groups, continue in many cities to *build up* densities."[25] Both Bauer and Kelly called for a new vision of urban renewal and redevelopment—one in which the construction of new public housing following demolition in urban cores would complement dispersal policies promoting suburbanization. Like Augur, Bauer was optimistic that these plans and the social welfare benefits that would follow could be achieved with little additional expense.

Bauer was not the only advocate to play the urban-social-welfare card in the defensive dispersal debates. Robert Moses (head of the New York City Slum Clearance Committee from 1948) also linked dispersal to an improved quality of urban life, although he disagreed about how far from the central city slum populations should be scattered. Evacuating these populations to suburban areas seemed to him "undesirable and unworkable," and so in a memo to Mayor William O'Dwyer criticizing how other civil defense planners had framed their task, Moses offered a counterproposal: that evacuation sites be constructed on heretofore unbuilt lands closer to the city center.[26] Within a year, Moses suggested, the city could build forty thousand units of low-income housing and relocate families from the "most congested slum areas" to Staten Island or the Bronx. Moses's longstanding agenda to demolish Manhattan's slums and to relocate slum dwellers found new urgency and new support in his call for defensive dispersal.

In the linkages they made between planning for urban security and planning for urban problem solving, these and other city planners moved the dispersal discussion in a direction that would establish a pattern for later collaborations among military and urban professionals.[27] National committees, commissioned research studies, and planning advisory boards were created for federal, state, and local authorities. These developments offered military and urban experts opportunities to create social networks, circulate among professional communities, and increase their prestige. Yet the history of intellectual and political support for dispersal planning was not matched by a history of equally vigorous action on the ground. While the planning process had a major impact on its participants' social status, the implementation of defensive dispersal failed to meet the specifics of these experts' recommendations.

From Argument to Action

Proponents of defensive dispersal compared city planning to weaponry, calling for a strategic mobilization for "space and decongestion" on the order of a military operation.[28] Yet cities are not military battalions. Federal, state, and local programs to push dispersal lacked the authority of military commanders over their troops. Over time, a variety of other forces, including private developers, federal housing policies, and improved road and highway networks did indeed lead to a significant dispersal of population and industry in the form of what we now call "urban sprawl." Yet this outcome did not proceed at the speed or match the specifications required by any master dispersal plans. In other words, dispersal happened, but not for reasons of enhancing national defense. The practical consequence of the defensive dispersal movement was a different kind of mobilization, one in which individuals and institutions were conscripted toward a new form of national service, the military-industrial-academic complex, creating a network that would long outlast the dispersal conversation.

Based on the calls to action from some of the nation's leading scholars and opinion makers, in the movement's early years the possibility of actually redesigning American cities looked promising. Federal efforts to promote dispersal of populations and industries began with a series of studies. From the late-1940s, the U.S. Office of Civilian Defense (later the Federal Civil Defense Administration, or FCDA), the NSRB, the DoD, the U.S. Strategic Bombing Survey, and the Atomic Energy Commission (AEC) were among the federal agencies that sponsored research on defensive city planning with the goal of developing specific dispersal recommendations linked to the power of existing enemy weaponry. For example, during this period the NSRB invited Tracy Augur to create a specific dispersal plan for federal facilities in Washington, D.C. The AEC released its own report describing the consequences of atomic attack on the city of Washington, D.C., with recommendations for civil defense planning. And the FCDA prepared a list of 185 cities it deemed Critical Target Areas and urged them to prepare dispersal plans.[29]

At the encouragement of federal officials, states and cities called in consultants to study the local situation and develop dispersal plans for their populations and industries. New York State and Massachusetts, for example, undertook area studies, with New York proposing to create a state agency for developing communities in nonurban areas. Cities including Baltimore, Chicago, Washington, D.C., and Denver, among others, embarked on their own area studies, and some even published local civil defense newsletters. Milwaukee—a leader in civil defense planning, with Mayor Frank Zeidler heading the American Municipal Association's Civil Defense Committee—invited Oscar Sutermeister, who later would serve as a consultant on Project East River, to assist in preparing a plan for a satellite community. Los Angeles studied the issue and concluded that the city already was sufficiently dispersed according to NSRB recommendations. At the request of NSRB, cities from Seattle to Chicago and Washington, D.C., simulated explosions to assess the effectiveness of existing civil defense infrastructure.[30]

But not all civic leaders were so enthusiastic. Civil defense officials in the nation's most concentrated urban area, New York City, were less receptive to the dispersal ideal. C. McKim Norton, a long-time advocate of dispersal in his role as president of the Regional Plan Association of New York and later as a consultant on Project East River, led industrial dispersal planning in the New York City area. Yet even though the city's planning commission chair acknowledged that "New York City is potentially a Number One target, if war envelops the globe," documents in the city archives show minimal interest in a comprehensive dispersal program.[31] The official line from the city's Office of Civil Defense stressed that whatever the nation's espoused urban decentralization or evacuation plans, at least half the residents of most large cities would have to commit to staying in order to keep industry and government up and running.[32]

Even for those states and cities approving dispersal plans, strategic recommendations often lacked the power of a strong enforcement mechanism. This certainly was the case for Washington, D.C. Despite the fact that since 1878 the District of Columbia's planning commission had been headed by an army brigadier general, despite the fact that Augur had developed a specific plan, despite the fact that government hearings had been held on the topic, despite the fact that government policy was encouraging dispersion, and despite the fact that eighty-five sites had been set aside for federal agency relocation between thirty and three hundred miles from the city, most federal agencies remained close to downtown because Congress declined to appropriate the $190 million necessary to implement the plan.[33] In 1950, the NSRB released a report that listed—with some concern—states' limited actions to date on civil defense planning. As a stimulus to keep this process moving forward, the NSRB created an idealized civil defense strategy for a fictitious state, "Columbia," with instructions for all other states and regions to follow suit. The American Municipal Association issued a similar report in 1958 with a discussion of defense planning at the city level based on surveys of mayors and city managers in 153 cities. This report, like so many others, suggested civil defense was a program with far more talk than action.[34] Civil defense planning remained stuck at the "planning" phase.

Industrial Dispersal

While planning for the survival of civilian populations was an important component of civil defense efforts, survival of key industries had its own explicit focus in public policy. Discussions and studies through the late-1940s in 1951 led to the creation of a federal industrial dispersal policy. Just as the Federal Civil Defense Administration had identified its broad urban target areas, the National Industrial Dispersion Program defined its industry-specific target areas (they were not the same). A year later, in August 1952, tax-amortization privileges were granted to defense-related facilities located at least ten miles away from population and industrial centers. These policies provided clear financial incentives to enhance dispersal for the nation's war-related industries, with the additional benefit of putting more factories in more congressional districts.[35]

As the FCDA had done for the general purposes of dispersal planning, the U.S. Department of Commerce asked city chambers of commerce to create "industrial dispersion committees" to help implement the policies by approving business location plans, indicating that these organizations had a continuing role in city planning. To enhance federal oversight, the department's Office of Area Development held several meetings on dispersal progress during the decade and requested that each committee submit overview studies of its metropolitan area.[36] By 1953, eighty-two areas had formed industrial dispersion groups, and twenty-seven dispersion reports had been approved. By 1957, sixty reports had been submitted (the reporting cities included Chicago, New York, Dayton, Pittsburgh, and San Francisco).[37]

However, in much the same way that civil defense studies commissioned by government and private sponsors in many cases were unable to proceed from plan to implementation, the creation of industrial dispersal committees was not the same as actually dispersing industries. Analysts from the University of Maryland observed that "dispersal standards were generously relaxed or totally waived" for many defense contractors; firms were "rewarded with tax privileges and classified as conforming to dispersal standards" even when this was not in fact the case.[38] Critics blamed such practices on a lack of funding to administer the industrial dispersal program, as well as a lack of central coordination in the nation's civil defense program. The industrial dispersal program offered neither stringent requirements with oversight nor appealing economic incentives for voluntary actions by city managers to promote the dispersal measures that security experts had specified. For example, municipalities would lose valuable revenues if industries relocated outside cities' taxable boundaries. Thus a 1956 evaluation study concluded that the dispersal program "has essentially existed in name only."[39]

Like the city planners who employed "national defense" as a rhetorical tool to advance their older goal of master planning the American landscape to create a network of garden cities, business leaders found defensive dispersal a good public relations tool to justify moves undertaken for other reasons. A number of prominent corporations did disperse some industrial and commercial locations. For example, industry giants such as IBM and Standard Oil moved their headquarters, and the Life Insurance Association of America promoted dispersal in its public materials. The Office of Defense Mobilization praised the actions of several military contractors in a 1957 report, observing that "companies which have made dispersion a major element in site selection included the General Electric Company, Glenn L. Martin Company, United Aircraft, Minneapolis-Honeywell and Lockheed Aircraft Company."[40]

Yet it does not appear that federal encouragement of dispersal was the primary motivating factor behind such relocations. For example, while General Electric was held up as a leader in industrial dispersal, Phil Reed, chair of the company board, was quoted by an NSRB spokesperson as saying the company chose dispersal to less-urbanized locations for economic reasons. In Reed's words, "Despite the fact that our decentralization has been fundamentally economic in nature, it is nevertheless true that it automatically offers many of the security advantages which have been advocated by the National Security Resources Boards, and in this respect is doubly desirable."[41]

Efforts continued throughout the 1950s to maintain connections between defense experts and urban leaders, combining military and civilian concerns. The Federal Civil Defense Administration brought mayors and local chambers of commerce on board by hosting numerous meetings on civil defense. For example, at a 1956 meeting of the Washington Conference of Mayors, the assembled audience heard from the nation's most senior defense officials, including the chairman of the Joint Chiefs of Staff, the secretary of state, and other senior defense advisers.[42] The American Institute of Planners developed a close working relationship with the NSRB, whose head occupied a place on the National Security Council. Special AIP committees discussed civil defense every year between 1953 and 1960. Yet even in cities whose official policies embraced dispersal, implementation of policies and plans continued to proceed slowly. In June 1956, the Office of Defense Mobilization sent action proposals to governors and mayors of cities with populations of more than fifty thousand, urging them to get moving. The following year, the Federal Civil Defense Administration prepared an operations plan for urban defense, with numerous maps suggesting the potential damage of an attack to the city of "Battleground, USA."[43]

Throughout the 1950s, proponents of defensive dispersal such as Coleman Woodbury and Catherine Bauer debated the extent to which urban dispersal was actually a coherent, tightly organized, and efficient federal program. Woodbury's urban redevelopment study identified how in fact local industries had been decentralizing since the 1890s, an idea Louis Wirth also had raised in his presentation to the American Municipal Association.[44] Both men pointed out that it was not obvious that dispersal efforts were so much a break from the past as they were continuous with it. Other observers noted the extent to which dispersal was already under way, thanks to uncoordinated initiatives including Federal Housing Administration loans, increasingly widespread highway development, and the razing of central cities. Whatever the exact origins and nature of urban dispersal as it was actually occurring in the 1950s, it is clear in retrospect that the defenses afforded by a decentralizing American landscape did not meet the stringent requirements of a master military plan.

The Twilight of Dispersal Planning

Despite the vigorous rhetorical and institutional support for dispersal policies, proponents of defensive dispersal whose primary concern was engineering a landscape suitable to defend against the technical capabilities of enemy weapons remained dissatisfied. Committees were formed, studies were commissioned, and recommendations were issued, yet the realities of program implementation were weak. In 1953, the same year as Bauer's proposal to link dispersal and social welfare planning and the same year as the AIP's declaration to serve national defense needs, Burnham Kelly offered a more skeptical view of what was being accomplished. Kelly extended Augur's argument to suggest the nation's intellectual capital was at risk unless civil defense considerations were incorporated more effectively into urban redesign. Two-thirds of the U.S. population was now living in cities, and that percentage climbed even higher for "skilled labor, technicians, scientists, management personnel, and other key men."[45]

Echoing Augur's earlier analogy to the principles of deterrence, Kelly compared dispersal of cities to the build-up of nuclear arms. He argued that the federal government needed to make a bolder statement to mobilize the nation. He pointed out that, while the National Security Resources Board had done a good job getting dispersion off the ground, federal industrial dispersion policy to date had been inadequate. Kelly cited other existing policies promoting decentralized development, such as the federal highway program and housing policies that had sparked a suburban housing boom, but he noted their standards did not reflect national security needs. In other words, if the suburbanizing nation was in many ways dispersed, this dispersal did not match the rigor of any master defensive dispersal plan.

Kelly was not the only public figure to argue that early steps to wed military needs with urban planning needed substantial strengthening in their implementation. For example, Richard Bolling, a Democratic congressman from Kansas City, in September 1951 authored an article on the politics of dispersal in the *Bulletin of the Atomic Scientists*. Similarly, Lewis Anthony Dexter, an industrial analyst at MIT and former adviser to the Democratic National Committee, observed that Project East River should have ignited a much more vigorous public discussion than it did. At a 1954 meeting of industrial dispersion committees, participants discussed the extremely slow progress of dispersal in New York City. Senator Hubert Humphrey was among those who observed that the National Industrial Policy (another name for the National Industrial Dispersion Program) had not been successfully implemented, and he argued for the creation of a federal "urban decentralization authority" to speed the process.[46]

Several years later, in 1960, Philip Clayton would complain that, still, too little had been done to convert plans into reality in the United States, whereas Russia had built 118 cities since 1951 with civil defense in mind. Clayton, the senior planner in the Comprehensive Planning Division of the Baltimore County Office of Planning and Zoning, observed in the *Journal of the American Institute of Planners* that incentive policies on the books continued to lack any serious power of enforcement.[47] Fundamental distinctions between military and civilian organizations, in particular the contrasting relationships between military commanders and troops versus city managers and urban populations, suggest why it was so difficult to coerce city agencies, housing developers, industries, and individual citizens to choreograph their movements with the kind of speed and according to the exacting specifications required of a master military plan.

Critics in the late-1950s, looking back on more than a decade of cooperation among defense experts and urban professionals, continued to express anxieties that rhetorical agreement was not translating into concrete results. Citing reports from the RAND Corporation on fallout problems and outlining the size of damage zones, Clayton and others wrote that they hoped the 1960s would be different. Yet the late-1950s marked the peak of linkage between defensive dispersal planning and urban development theory. The 1960 conference of the American Institute of Planners included a session on planning and nuclear warfare, yet discussions about dispersal began to slow soon after. The declassification of information on the latest weapons capabilities made it clear to wider audiences that nuclear weaponry had advanced to a point that all citizens were possible targets. (This information was already familiar territory to the defense experts who worked with classified information; analysts at RAND who had access to classified data detailing the capabilities of weapons of mass destruction were among the earliest skeptics that dispersal would actually offer useful defense.) As a 1961 article in the *Bulletin of the Atomic Scientists* explained, "What would people in cities do, even if their city were spared a direct hit? In most cases, they would die."[48] Historian Kenneth Rose confirms that, by the 1960s, national and local politicians saw more value in active defenses, such as antiballistic missiles, than in passive forms of defense such as dispersal planning, highways, and bomb shelters. In light of the knowledge that the Soviet Union was preparing an antiballistic missile system, pressure mounted for the United States to create a similar system. In 1967, Defense Secretary Robert McNamara proposed the development of an antiballistic missile system to be known as the Sentinel system.[49]

The fifteen-year engagement of defense experts, urban leaders, and the federal government around the topic of dispersal did not have obvious effects on the physical landscape. For average urban dwellers, the dispersal movement's effects were negligible. For the "power elite," by contrast, this engagement produced a marriage with lasting effects. Conversations about dispersal, by bringing together atomic scientists, military strategists, professional planners, business interests, and federal and local governments produced important social networks that would shape physical and social planning initiatives in the coming decades. The common ground these interest groups found in simultaneous efforts to promote national security and urban social welfare set a tone for future collaborations.

In 1953, Henry S. Churchill, a chief planner at Eastwick Planners in New York City, precisely forecast the direction these collaborations would take. Every planner, he urged, must read Norbert Wiener's *The Human Use of Human Beings*. The technical and technological innovations Wiener described could be the end to urban problems—if adopted for such humane ends. "We are at the commencement of a crucial period in city development," Churchill declared, "and the technician is an important, if not a decisive, factor in the turn of events." With so much power vested in the hands of technicians, a major decision confronted U.S. planners: to choose "between the roads that lead to the City of Man or the City of Eniak."[50] Although he misspelled the acronym for ENIAC, America's first electronic digital computer, Churchill raised an important point. From atomic energy to computers, advances in technology based on military innovation created a crossroads for U.S. city developers as they entered the era of "postwar planning." The Janus-faced techniques and technologies spawned by World War II—the focus of Wiener's book—might continue to be associated with destructive purposes. Yet they might also have more humane possibilities.

In their proposals to reshape the physical landscape, proponents of dispersal had identified one opportunity for military expertise to find its way into city planning and management, a means to promote their ideas about postwar planning and simultaneously contribute to the national defense. Yet as conversations about postwar planning continued, military techniques and technologies infiltrated expert thinking about urban development in other, less visible, ways. A new type of urban expert—the "technician," or technocratic defense intellectual—began to emerge. In Churchill's words, his appearance was "an important, if not a decisive, factor in the turn of events."

As the 1960s arrived, it became increasingly clear to urban planners and managers that middle-class, white citizens who could afford to do so were continuing and indeed accelerating their move to the suburbs. Nevertheless, urban agglomerations remained. It is in those urban spaces, where many of America's poor, immigrants, and minorities continued to live, that the next part of this story unfolds. Together, the following chapters investigate cold war urban history in light of these demographic changes, focusing on how a range of technical and technological products of wartime defense research found new uses in urban planning and management. The potential benefits of adopting these tools displaced dispersal as a primary conversation topic as defense experts offered urban leadership a new challenge: could they cooperatively use military innovations to improve the management of American cities?

Part I / Command, Control, and Community

CHAPTER TWO

The City as a Communication System

In "How U.S. Cities Can Prepare for Atomic War," a 1950 article in *Life Magazine,* Norbert Wiener, a professor of mathematics at MIT, joined the dispersal conversation. Wiener expressed his fear that centralized American cities—difficult to evacuate and difficult to defend—were easy targets for a nuclear strike. This father of cybernetics proposed accelerating the trend toward suburban growth as a defense strategy. Highways—"life belts," in his terms—would serve as conduits away from the city in the event of a nuclear attack. Wiener argued that dispersing the population, a "long overdue reform," would simultaneously alleviate urban problems.[1]

Wiener was not the first to suggest that U.S. cities were likely targets for attack, nor that dispersal would solve urban problems. Yet his cybernetic view of cities as communications systems offered a new rationale for the plan. "A city is primarily a communications center," he explained, "serving the same purpose as a nerve center in the body." Cities functioned best when information could easily be exchanged, and the persistent "traffic jams in streets and subways" signaled these exchanges could be much improved.[2] With the basic principles of cybernetics suggesting that "the distinction between material transportation and message transportation is not in any theoretical sense permanent and unbridgeable," Wiener argued that communications technology could knit together a physically dispersed population.[3] In the near future, he predicted, transportation of increasingly sophisticated materials via communications networks would become common. Wiener's vision of a distributed population alongside a sophisticated information network was the centerpiece of his vision for a healthy and humane future society.

Wiener died in 1964, before ARPANET would make his predictions about message transmission a reality. Yet during the last years of his life, several urban experiments got under way to bring cybernetic principles to city planning and management. Communication technologies would not be used for dispersed citizen-citizen or citizen-government communications, as Wiener imagined. Rather, his image of cities as information processing systems would be applied to reshaping urban planning and management practices.

In every era, one or two "images of the city" dominate urban planning and management; techniques and technologies can play a defining role. In his bestseller *The Image of the City* (1960), Wiener's MIT colleague Kevin Lynch, professor of city and regional planning, investigated how average citizens experience cities.[4] The book—an effort to aid designers struggling to improve the urban order and make it more responsive to users' needs—sought to understand how city residents make mental maps of their everyday environment. By juxtaposing laypersons' images with expert images from colleagues in the field, Lynch revealed a fundamental disconnect between the goals of experts and the objectives of average citizens. His book was a metaphor for the failures of urban renewal. Lynch's findings made clear that the image of the city most commonly put forth by experts, which had driven so much of urban policy, planning, and management in the 1950s, would need significant revisions to better serve city populations.

For urban decision makers seeking a new "image of the city" in the wake of urban renewal's failures, Wiener's conception of cities in cybernetic terms held wide appeal. In the late-1950s and through the 1960s, the belief that cities might be understood as communication systems gained popularity among both academics and practitioners.[5] From this vision, it soon followed that principles from cybernetics and computing technologies found broad application in city planning and management. By 1966, Richard Meier and Richard Duke would write in the *Journal of the American Institute of Planners* that the urban professions were experiencing a "revolution" in theory and practice, with computers at the center of this transformation.[6]

The image of the city as a communication system drew upon the science of cybernetics and its close cousin systems analysis, a technique for military decision making developed at the RAND Corporation in the 1940s. Electronic computers, first developed for ballistics-data processing, were also military innovations of the 1940s. That the techniques and technologies adopted to transform city planning and management in the postrenewal era were first developed for the task of military planning and management suggests another kind of revolution was under way in the American urban professions, one in which the military-industrial complex played a starring role.

The Origins and Early Military Applications of Cybernetics

The science of cybernetics is most often associated with the work of Norbert Wiener. During World War II, Wiener, like many of his colleagues, was tapped by the armed forces to collaborate on a military project. In Wiener's case, it was gunfire control. While doing research to develop range finders for antiaircraft guns, Wiener became interested in machine learning. Servomechanisms enabled each gun to predict an airplane's trajectory by making use of information about previous trajectories. The concept that inanimate machine systems could learn from past events and use this information to predict the future intrigued him. Collaborations with his MIT colleague Julian Bigelow, an engineer, and the Harvard neurophysiologist Arturo Rosenblueth helped Wiener to develop cybernetics, a science based on the assumption of consonances between living organisms and machines.[7] In cybernetics, humans, machines, and organizations are systems of communication and control.

Cybernetics is a science of systems; as an interdisciplinary science, it merges concerns and tools of physical and biological sciences to analyze and understand "control in the homeostatic or self-regulatory sense rather than in the coercive sense."[8] Cybernetics is a science emphasizing dynamic processes, action and reaction. Early theorists sought to understand mechanisms of feedback and self-regulation in a variety of types of systems—how they work—in order to simulate, and in some cases to manage, these systems.

A series of conferences between 1944 and 1953 on "circular causal feedback mechanisms in biological and social systems" further developed the theory of cybernetics. The Macy Conferences (as they were named for the sponsoring Josiah Macy Foundation) brought together researchers across disciplines— electrical engineers, mathematicians, physiologists, anthropologists, and mechanical engineers—to discuss the emerging science.[9] Pioneers in the field include Herbert Simon, whose research ranged from economics to public administration to artificial intelligence, and Claude Shannon, an MIT-trained mathematician who helped to develop related ideas in a field called information theory. But most prominently, Wiener put the public face on cybernetics. Wiener's best-selling books *The Human Use of Human Beings* (1950) and *God and Golem* (1964) brought his ideas to broad audiences. It was in these best-sellers that Wiener coined what is the contemporary definition for cybernetics: the field of "control and communication theory whether in animals or machines." This wording conveyed the notion that communication among subsystems for purposes of control was the key to organismic life. So, too, society could "only be understood through a study of the messages and the communications facilities which belong to it."[10]

As cyberneticists sought applications for their theories, they generated interest across a diverse array of research fields. One branch of support came from physiology and biology, concerned with living things. Katherine Hayles has described the influence of cybernetics on the neurosciences, in, for example, studies of neural networks. Lily Kay has documented molecular biology's transformation during the 1950s to represent itself as a communication science. Evelyn Fox Keller has described how, since the 1960s, the metaphors that developmental biologists use to describe organic phenomena have borrowed from the cybernetic language of feedback and systems.[11]

Support for applying cybernetic principles to research on nonliving systems emerged from organizations, academic and nonacademic, studying management, engineering, and control. RAND and MIT stood at the forefront of this trend. With their heritage of mathematical innovation and ties to the armed forces (long interested in communication and control), these and cognate institutions offered ideal laboratories to transform cybernetic principles into management practices. At RAND in particular, during the late-1940s and early-1950s, principles from cybernetics were synthesized with other fields and reconstituted as systems analysis. This synthesis created an applied interdisciplinary endeavor that, according to historian David Hounshell, integrated several branches of math and science, including game theory, probability and statistics, econometrics, and operations research.[12]

Systems analysis is based on the idea that all problems can be understood in the context of systems—with human, machine, and organizational components. Wiener's research on the use of antiaircraft guns during World War II is one example; Wiener attacked the problem by considering the human operator as part of the total machine's system. Combining this cybernetic view with mathematical analysis, and later with computer simulation and modeling, systems analysis grew to become the premier science of military decision making in the cold war.

The military has a long history of using models for strategic decision making. Many such models were conceptualized as "war games," and these games date back several centuries, at least to the Prussian army. During the late-1940s, RAND, the U.S. Air Force think tank, became a central site for cold war gaming and a place for further refinement of the systems analysis technique to evaluate decisions for wartime scenarios in a nuclear age. RAND mathematicians—among them Herman Kahn—drove much of the nuclear strategy. In RAND strategy games, players reduced complex political contingencies to mathematical formulae as aids to decision making.

As the technique developed further at RAND, systems analysts could quantify increasing numbers of variables and then use this quantification, combined with mathematical modeling, to compare alternative solutions to problems under conditions of uncertainty. They accomplished this by running simulations of the consequences of specific decisions. Expressing complex political categories as mathematical equations allowed them to be introduced into a computer. (Electronic computers had been created only a few years earlier to speed the processing of ballistics data, and during this early period they were still closely identified with military applications.) Computers were able to engage more effectively and more quickly than ever before with more variables. The explicit goal of this marriage of technique and technology was to make decisions with the highest likelihood of a positive outcome, however leaders chose to define that outcome.

RAND's ties to the aerospace industry (its origins were at Douglas Aircraft and in contract work for the U.S. Air Force) and to elite universities (from which it drew many of its consultants), dispersed such simulation games beyond its walls. Aerospace applications quickly followed military ones, and at several institutions (the Jet Propulsion Lab, Boeing, Lockheed, and TRW) systems analysis became standard management practice. Following quickly from RAND's military-focused exercises, MIT, also a wartime leader in contract military research, ran a number of strategy games focused on political and military affairs. Faculty expanded applications of gaming to topics of conflict and diplomacy both abroad and at home. For example, Professor Ithiel de Sola Pool, who directed communication research at MIT's Center for International Studies (CENIS) and later helped to found the Department of Political Science, became a creative early user of the tools. He took the uses of simulations for political gaming in a new direction, to model voter behavior in a project he called "Simulmatics." Among his other early simulations were COMCOM (under development from 1962) and CRISISCOM (under development from 1964). The former was about communication structures in Communist countries and the latter about decision makers' perceptions in crisis situations.[13]

Beginning in the 1950s, alongside uses for decision-making games, military and political leaders sought to create computer-based systems for real-time purposes of national defense. The same institutions that had helped to develop both the theory of cybernetics and systems analysis and their early applications in strategy games, RAND and MIT, would go on to play key roles in developing new defense applications during the 1950s and 1960s. Major examples of such technological innovations included the Strategic Air Command Control System, the NATO Air Defense Ground Environment, and the World Wide Military Command and Control System. Each was a command-and-control system centered around computers. Each instantiated Wiener's principles about the informational basis of national security, wherein "information is more a matter of process than of storage."[14] According to this view, national security was based not on vast storehouses of secret knowledge recorded and stored in books but rather on constantly updated, real-time information.

The SAGE air defense system is the premier example of a new tool that rose to meet the challenge Wiener offered (SAGE is an acronym for Semi-Automatic Ground Environment). Designed during the late-1950s, the system depended upon a combination of computers and radar to defend the United States against bomb attacks. Its task was to use radar to intercept hostile aircraft and to jam their autopilot systems. Mathematical modeling combined with real-time information processing made it possible continually to update the database with incoming information. Based on these data, SAGE could then make predictive decisions about where enemy aircraft would be at future points in time.

SAGE combined techniques, technologies, and expertise from several high-profile institutions, including the Lincoln Laboratories, a spinoff of MIT and SDC (System Development Corporation, itself a spinoff from RAND, created to assist the Air Defense Command of the Air Force). The technical collaborations that led to the creation of the machine also spawned the MITRE Corporation (an acronym for MIT Research and Engineering), a think tank comprised of much of the staff of Lincoln Labs' Division 6. Jay Forrester, who had worked at MIT developing the Whirlwind computer for the U.S. Navy during World War II (with its first ever random-access core memory), went on to play a leading role on the SAGE project as head of the digital computer division of Lincoln Labs. Paul Edwards has described Forrester's close ties to the military and how his participation with SAGE helped to make it the first system to use computers for control, rather than simply for information processing.[15] Yet SAGE was never actually tested in a military conflict; despite the fanfare, questions about its effectiveness persisted.

Applications in Vietnam

SAGE never had the opportunity to prove itself in war, but related techniques and technologies did. Systems analysis and computer simulations were essential components of military operations during the Vietnam War. Historians emphasize the pivotal influence of Robert S. McNamara, former president of the Ford Motor Company, whom President Kennedy appointed in 1961 to head the U.S. Department of Defense. McNamara brought RAND staff, including Charles Hitch and Alain Enthoven, to DoD to implement a "planning, programming, budgeting system" (PPBS; also referred to as PPB), a new technique for cost-benefit analysis derived from systems analysis. PPBS became the brand name for the style of systems analysis created in the Department of Defense in 1961 and 1962.

While some of the roots of systems analysis clearly lie in military projects, evidence suggests that when first brought to the DoD as a budgeting tool, PPBS was the subject of some consternation among military commanders. DoD under McNamara had imported a team of economists and statisticians with Ph.D.'s, not years of military service, to run analyses of military budgets, and with this tool shifted standards in goal setting from concurrent development, which emphasized parallel processing and rapid weapon development, to phased planning, a more sequential method that focused on budgetary constraints and a more top-down form of planning to which PPBS was well suited. Yet eventually, the tool became identified with military analysis.[16]

Alongside the importation of PPBS, McNamara created an Office of Systems Analysis in the Department of Defense, staffed by civilians, with the job of overseeing strategic planning and weapons programs. While historians such as Martin Van Creveld have argued that the Office of Systems Analysis (later the Office of Program Analysis and Evaluation) was not a major player in the Vietnam War, they nonetheless observe that the systems approach, especially the kinds of informational inputs required, created the "information pathologies that characterized the war in Vietnam and made no small contribution to its outcome."[17] As part of a war effort that included social, political and economic interventions alongside traditional military combat, information systems that measured and analyzed social, political, and economic data on the ground became important aids to strategic decision making.

During the Vietnam conflict, McNamara wanted to know what was actually happening in individual hamlets and provinces overseas. DoD therefore initiated several social and political information-gathering programs and information system development projects, based on the belief that this information, properly quantified and analyzed, would point decision makers toward the military strategy that best matched their aims. As part of these information-gathering efforts, RAND staff were contracted to conduct several analyses of the Vietnamese population, including an assessment of the pacification program. For example, J. A. Wilson compiled geographic and population data for the region. David Elliott and William Stewart prepared an interview-based field study to understand the effects of the pacification program following the arrival of U.S. forces; like other political analyses of the conflict, their report described the political "system" in cybernetic terms.[18]

As part of information system development efforts, a Hamlet Evaluation System (HES, also referred to as Hamlet Evaluation Survey) was implemented in January 1967 to give geographically specific feedback on pacification efforts and their outcomes in individual villages in the Vietnamese countryside. The HES represented a synthesis of systems analysis and computing as a management tool. Adopting a systems approach to their study of political stability and instability, civilian and military analysts in the field gathered and then quantified information on the status of security and development in individual hamlets.[19] Questions about the political climate and social interactions were scored on a scale of one to four, making it possible to introduce them into computers.[20] Once these reports were computerized and compiled, they were sent on to Washington, D.C., for analysis. For a broader public, monthly Southeast Asia analysis reports were published, fifty in total between January 1967 and January 1972, tabulating details of the military, political, economic, and social situation in each of many areas. According to Robert Komer, who headed psychological warfare operations in the pacification program as the deputy for CORDS (Civil Operations and Revolutionary Development Support) to U.S. Army General William E. Westmoreland, these "analyses often had direct impact on the Washington decision process, notably on issues of whether the attrition strategy could succeed, on the impact of our bombing, and on how US withdrawals might best be conducted."[21]

Nevertheless, Thomas Thayer, who served as director of the Southeast Asia Division of the Office of the Assistant Secretary of Defense for Systems Analysis, was aware of the Hamlet Evaluation System's many imperfections. Users launched repeated efforts to test its validity and make improvements in order to create a more "objective" survey of conditions in the field. In one such effort, the Simulmatics Corporation, staffed largely by social scientists—indeed, created by de Sola Pool, whose studies of U.S. public opinion and voter behavior comprised his first Simulmatics project—was invited to conduct a field study to evaluate the HES by comparing it to more qualitative interview-based data.[22]

According to Colonel Erwin Brigham, chief of the CORDS Research and Analysis Division at the Military Assistance Command Headquarters in Vietnam, the HES continued to find new applications as the systems developed and improved. This included early uses of the information gathered as input data for early geographic information systems. One of these early innovations was the Province Hamlet Plot, an overlay that quickly came to be used in operations by military officers from both the United States and the Republic of Korea, as well as by the national police force of the government of Vietnam.[23] Despite its imperfections, then, the HES represented an advance in military mapping and information system development. Able to chart change in the status of individual hamlets and display information on a computer-generated map, it instantiated the principles of the dynamic national security system that Wiener had theorized about in his 1950 book.

HES was only one of many such geographic information systems created and used during the Vietnam conflict. Among the related tools developed for the war were an Operations Analysis System (1963–73), a management tool with subfiles including an Enemy Base Area File (1966–71), a Southeast Asia Friendly Forces File (1966–72), a Terrorist Incident Reporting System (1967–73), a Viet Cong Initiated Incident File (1963–71), and a Vietnam Data Base (1963–71). The overall system, which had computer mapping capabilities, interlinked these smaller data files, making it possible to abstract answers to numerous questions and then to map them for a graphical view of the status of the conflict in order to make decisions about next steps. One question that could not be answered, however, was, "When will the United States win the war?" An apocryphal story recounts that one of the DoD's computers was asked in 1968 when America would win the war. The computer responded that America already had won—four years earlier, in 1964. This error foreshadowed future criticisms of the role of systems analysis and computers in the war's final outcomes.[24]

It is important to stress that, like the users of the HES, many early players of simulation games appreciated that their models were simply models, imperfect and oversimplified representations of the real world. Systems analysis and computer simulations were the subject of much constructive criticism in their early forms at RAND. John Raser wrote of RAND strategy games that participants understood at the time "the simplifications imposed in order to permit quantification made the game of doubtful value for the assessment of political strategies and tactics in the real world."[25]

Nevertheless, as talk of practical applications became more serious, hopes were high that these new systems were more "objective" tools for decision making than their predecessors and that their validity could be improved. Some proponents disregarded early criticisms, arguing that simulations were value free and that that very neutrality was a central motivation for adopting them. William Lucianovic, a professor of public management sciences at Case Western, was one such enthusiast, reporting that "games eliminate any side motives that a politician, businessman, or administration might have."[26]

Others believed optimistically that advances in science and technology might lead to future models and decision-making systems that would mirror reality more closely. So, for example, at a 1969 conference at the American Academy of Political and Social Sciences, C. West Churchman, who worked on SAGE (and went on to direct research at SDC in the early 1960s and later the space science lab at the University of California, Berkeley) imagined a futuristic public policy scenario for the year 1990. In his scenario, the computer modeling of war games and the real-time information of a system such as SAGE offered an opportunity to create the ultimate decision-making technology. Future wars would find the U.S. president simply asking his computer about the likely consequences of specific actions and the computer providing guidance. The descendants of existing systems would not only anticipate the course of specific actions but also suggest alternative military and political strategies.[27] The hope was that systems analysis, in partnership with computing, would improve real-life military strategy. Military and political decision making would become objective and elite sciences.

Others suggested still another way to improve the public face of cybernetics: reduce the hype about what these tools could do. This was the view of RAND's Herman Kahn, who suggested that by 1968 "exaggerated claims" about benefits of the new methods were on the decline at the same time that systems analysts were improving the quality of their work. If the trend continued, he observed, "we may well come out with a match between claims and product."[28] Kahn's suggestions would prove unpopular.

A More Human Face—and a Larger Market

As with many of his colleagues in the physical sciences who worked to develop the atomic bomb, Wiener became disillusioned by the violent uses to which his science of communication and control was being applied. While he had participated in military-sponsored research during World War II, in the postwar period Wiener would plead for more humane uses for scientific knowledge. *The Human Use of Human Beings* (1950) is an eloquent expression of this view.

Wiener observed that his definition for the term *cybernetics,* and its early associations with military objectives, was neither its first nor its definitive meaning: "I found later that the word had already been used by Ampère with reference to political science, and had been introduced in another context by a Polish scientist, both uses dating from the earlier part of the nineteenth century."[29] In the nineteenth century, the word—which in Greek means steersman—was synonymous with governance. Might the governance of civil society once again become the focus of cybernetics? Beginning in the late-1950s, systems analysts—and an increasingly wide array of government leaders—began to speculate that if society could be viewed as a self-regulating organism or machine, one of the complex social systems to which cybernetic principles might be applied most productively was government. Eventually this view would encompass understanding, regulating, and troubleshooting the governance and operations of American cities.[30]

From the late-1950s, researchers from TRW (a systems engineering firm and major defense supplier), RAND, and other defense contractors began to publish in the *Journal of the American Institute of Planners* and *Public Administration Review,* suggesting how techniques and technologies from military operations research such as systems analysis and computer simulations might offer a new direction for city management. By the early-1960s, the pages of these journals, as well as the transcripts of conferences on urban planning, city management, and public administration more broadly, made frequent reference to systems analysis, cybernetics, operations research, and computers. This was both an exciting intellectual development and a highly practical plan. For while overall the cold war nurtured the expansion of America's military-industrial complex, a downturn in federal defense spending arrived in the early-1960s. This was due in part to reductions in fear of the Soviet threat (President John F. Kennedy signed a limited test-ban treaty) and in part to the projected conclusion of the Apollo Program. In this uncertain economic climate, executives and engineers from institutions such as RAND and SDC decided that the survivability of their organizations depended upon finding ways to transfer their innovations beyond military clients. New markets were sought, and city planning and management quickly emerged as targets of opportunity. Three explanations were widely repeated to support the fit between urban management practices and military management tools.

A first rationale for technology transfer was the idea that computer simulations were simply the latest generation of models for a profession long used to them. Both in military strategy and in planning and urban design, there is a long history of using models as aids for command-and-control-style, top-down decision making. While planners had little experience with mathematical simulations or computers before the late-1950s, the profession had long made use of models in the form of maps, drawings, and three-dimensional scale models. This variety of visualization techniques played an important role in professional practice, key for developing many practitioners' "image of the city," to use Kevin Lynch's phrase.

In a 1957 article in the *Journal of the American Institute of Planners,* M. C. Branch extended the analogy by comparing the work of planners and systems analysts. Branch, who had worked as a research assistant on President Roosevelt's National Resources Planning Board and was at that time employed by TRW, had experienced both military and urban management firsthand and had observed numerous similarities.[31] Several years later, an article by Ira Lowry at RAND suggested city planners could gain much by educating themselves to use the latest generation of military planning tools. Computers were not "wiser than their masters," he explained, but rather took on the most repetitive tasks with efficiency and accuracy.[32] Tracy Augur had employed the rhetoric of both past and future to promote urban dispersal, finding resonances in the "greatest planning of all time" and, simultaneously, suggesting how postwar planning in the atomic age had to break with its "accustomed modes of thought." Similarly, proponents of the new military management tools argued that these innovations offered relief from the monotonous tasks planners had always done and provided them with something completely new. In the words of one author of a review compendium, "urban information systems and the systems approach to planning are strongly interdependent, largely because of the information demands associated with the 'systems approach.'"[33] Computer simulations could handle more variables and visualize their interactions in new ways, with special relevance for confronting the urban problems that increasingly occupied center stage on the national agenda.

The consequences of adopting these tools would be a new focus for planners on "problem orientation." Problem orientation was the notion that older categories of understanding urban problems were limited and that a new way of thinking that transcended these categories was required. Problem orientation aimed to do this by framing urban problems as processes, focusing on how challenges such as urban blight were "in continual self-adjustment."[34] Adopting the cybernetic language of feedback, homeostasis, and control in their musings on urban processes, city planners began to argue that cities were cybernetic systems, making systems analysis and computer simulations the necessary tools for a problem-oriented approach to administrative challenges. This language choice helped these tools to become accepted standards in professional practice. Some took the analogy even further. In a 1960 presentation to the American Institute of Planners (a session on Session on Systems Analysis and Operations Research in Planning), Stanford Optner, one of the earliest contractors developing data processing systems for the city of Los Angeles, redefined the profession of planning itself in cybernetic terms. City planning would become a "control mechanism" to improve decision making in the city system because "only the planning department has the potential to develop all of the tools and resources necessary to solve the complex urban problems which exist, today."[35]

A second, oft-repeated justification for the adoption of military innovations for urban planning and management grew from this redefinition of city processes in cybernetic terms. The science of cybernetics built on analogies between humans, animals, machines, and organizations. In so doing, it could unify disparate traditions of planning and urban analysis based on understandings of cities as organic systems and as machines.

The quest for a coherent science of the city was not new. Observers of the urban order had long aspired to a scientific understanding of city processes. In nineteenth-century France, the city was analogized to a human body, and urban planners used the vocabulary of surgeons. In America during the 1920s and 1930s, cities were characterized as ecological systems, and sociologists applied models from the biological sciences to understand how cities function. In 1961, Jane Jacobs, in her classic critique of urban renewal *The Death and Life of Great American Cities,* looked to the biological sciences as a model for reforming urban planning.[36] Alongside visions of cities as organic systems were parallel interpretations focusing on the city, its economy, and the urban political system as a machine. This tradition was especially prevalent during the first three decades of the twentieth century (the "Machine Age"), when technocratic elites imagined the nation as a complex machine, fantasizing how it might be possible to engineer the political and social system, including cities, just as managers had increased the efficiency of factory machines.[37] Cybernetics offered the potential to embrace and merge understandings of the city based in biological and physical sciences with characterizations of the city aligned with mechanical systems and organizations. Richard Meier's *Communications Theory of Urban Growth* (1962) expresses this dual vision, describing cities simultaneously as complex living organisms and as machines.[38] According to this view, systems analysis and computing were "revolutionary" innovations in their ability to draw together multiple classic metaphorical "images of the city," knitting them into a vision of the city as a cybernetic system.

One appeal of this image of the city was that cybernetic theory captured action and feedback. By extension, only tools such as databases and computer simulations that could be updated—rather than maps or three-dimensional models—were equipped to truly represent and analyze urban processes. Just as Norbert Wiener had argued that national security should be based not on static knowledge recorded in books but on a process of constantly updated information, urban scholars began to argue that city planners needed to employ similarly dynamic tools. So if, in the words of Melvin Webber, a professor of city and regional planning at the University of California, Berkeley, the urban community is a "form in action," then a master plan should not be "a fixed and static document."[39] Rather, as Glenn Johnson, of the Los Angeles Department of City Planning, and Leland Swanson, of the University of Southern California (USC), would suggest in the sourcebook for their 1964 course "Cybernetics and Urban Analysis," a city plan must be "subject to restudy and revision from time to time, as required by changing conditions."[40]

Increasingly in the 1960s, professional definitions of cities and their problems incorporated language that in turn defined systems analysis and computers as ideal management tools. This especially occurred in discussions about urban development initiatives such as the Community Renewal Program (CRP), where city planners and managers argued that "the CRP process should operate continuously to guide renewal of the dynamic changing city environment so that, as virtually a living organism, the city will not die prematurely."[41]

A third justification for the use of military innovations in the city, widely repeated, was that it would help to transform urban planning, management, and decision making into scientific endeavors. In so doing, these approaches would offer the scientific solution to urban problems that had been missing from previous efforts at urban renewal. Americans' esteem for math, science, and engineering grew dramatically in the two decades following World War II. During this period, the U.S. government increased its sponsorship of scientific and technological research, publicly proclaiming their value for maintaining a strong national defense. Both large-scale federal funding for and an emphasis on linking research to defense priorities were defining characteristics of U.S. science and technology in the postwar period.

Historians of science and technology have examined the development of military-university contract research in this era and how the increased prestige for math and engineering in turn had important effects on the social "sciences." Disciplines from political science to sociology to geography to management tried to remake themselves into mathematically verifiable endeavors, both for their own prestige and to attract federal funding. School of design and public administration were no different. The idea of a scientific attack on urban physical and social problems held special appeal for both academics and practitioners because it seemed to promise a remedy to the profession's most recent failure—urban renewal.

Urban renewal was the main federal urban program of the 1950s, a large-scale attempt to fix urban troubles primarily through physical redesign. Interpreters pinned much of the blame for its lack of success to its excessively narrow focus on physical planning and its concomitant neglect of social planning. They criticized the fact that urban development projects were contracted one by one, foxhole by foxhole, with little thought to their interactions.[42] Increasingly, they pinned some blame on the lack of a scientific approach. Among the widely repeated criticisms of urban renewal was that the local effects of this large-scale federal program had not been obvious or quantifiable. In other words, despite heavy financial and institutional investments in problem solving, it wasn't evident what positive difference federal monies had made.[43]

This crisis of social measurement was widely documented, and it would come to a head when rioting erupted in several cities. In testimony before the Senate Labor Committee, President Johnson's chief adviser for domestic affairs, Joseph Califano, commented that because no nationwide records of welfare recipients existed, the administration had taken nearly two years merely to find out who the seven million people receiving about $4 billion annually in welfare payments were. Further, after rioting in Watts in 1965, federal investigators following up on those disturbances found a severe lack of available data about community residents and their living conditions. According to Califano, the ability of federal officials to assess and review the outcomes of large-scale investments in social programs "more nearly resembles the intuitive judgment of a benevolent tribal chief in remote Africa than the elaborate, sophisticated data with which the Secretary of Defense supports a major new weapons system."[44] This lack of evaluation data and of scientifically rigorous methods for its analysis became a source of widespread criticism and embarrassment for both federal and local officials. Califano used the contrast between resources available to the nation's defense and urban agencies to call for additional investments to bring scientific evaluation to America's social programs. Little did many supporters recognize how many of the tools the federal urban agency would adopt were products of the nation's military.

It is important to recognize that even before the escalation of urban violence in the mid-1960s, the perceived failures of urban renewal already had city planners and managers seeking to build new expertise and making efforts to recapture professional clout. Leaders in the field such as Lowdon Wingo, at Resources for the Future, and Robert Wood, at MIT and later at the Department of Housing and Urban Development (HUD), promoted the idea that information systems and systems analysis could respond directly to the failures of urban renewal.[45] A new conception of urban planning and management as rigorous, applied sciences might provide the solution that would help to manage complex data sets, depoliticize the political process, and force a scientifically verifiable outcome.

The effects on the urban professions of this turn toward math and science quickly became evident in the disciplines and universities that educated urban planners and managers. MIT opened an Urban Systems Laboratory, where affiliated faculty such as Jay Forrester—with no formal training in urban research—undertook studies of "urban dynamics." Public policy programs developed courses in decision sciences and new academic degrees in such areas as public administration science. By the mid-1960s, schools of urban planning and public administration were offering courses with titles such as "Systems Analysis and Urban Planning" (at the New School for Social Research, taught by RAND consultants), "Cybernetics and Urban Analysis" (at USC), and "Urban Gaming" (at the University of Pittsburgh), with lessons on quantifying "inputs," "outputs," and "feedback."[46] Cities in turn imported these scientific approaches for use in newly created agencies, such as the New York City Office of Management Science and the Los Angeles Community Analysis Bureau.

Federal Backing for Public-Sector Cybernetics

Efforts to transfer cybernetic techniques and computer technologies from military management applications to the civilian public sector date to the early-1960s, precisely the time that the technique was reaching its full flower as a military tool. President Lyndon Johnson was so taken with Defense Secretary Robert McNamara's uses of PPBS in the Department of Defense that in 1965 he issued an executive order insisting that all federal agencies use the method to measure the output of the programs they directed. He did this even though the benefits of PPBS-based decisions were as of yet unproven. In the words of Daniel Patrick Moynihan, PPBS "colonized the federal establishment."[47] Offices of Program Analysis and Evaluation were created across federal agencies, where they persist today. Henry Rowen, deputy assistant secretary of defense for international security affairs under McNamara in the early-1960s and later assistant director of the Bureau of the Budget and president of RAND, and Joseph Califano, who had worked in the Office of the Secretary of Defense from 1961 to 1964 as assistant to McNamara before becoming President Lyndon Johnson's special assistant for domestic affairs, both served as bridges in this transfer. Each recognized that a large number of federal agencies would need an army of systems analysts to implement the new management tools.[48]

Federal agencies responded differently to the executive order. While some were lukewarm, others, in particular HUD, offered an enthusiastic response. This enthusiasm was likely based on the fact that several cities (for example, Pittsburgh and San Francisco) were beginning some early experiments with similar analytic techniques, even though their methods were not branded formally as applications of PPBS. Robert C. Weaver, the secretary at HUD, and his undersecretary, Robert Wood (a former colleague of de Sola Pool at the MIT Department of Political Science), went on the lecture circuit to publicly express their hope for a successful transfer of defense tools to improve urban renewal programs. Wood's testimony to one audience explained that the federal department's enthusiasm was based on the "increasingly realistic assessments and prediction" these technologies offered. "Actual data" and the rational "scientific" method would replace "seats-of-pants judgment" and "political intuition."[49]

Both Weaver and Wood, like other participants in these discussions about urban experiments, were not naïve technological optimists; rather, they understood how technologies might be put to positive or negative uses. They acknowledged that analyzing and modeling cities inevitably would be more complicated than analyzing and modeling weapons systems. Nevertheless, Weaver and Wood urged city leaders and systems experts to make efforts to test military management tools in urban contexts. For example, at a June 1966 conference at the National Academy of Science's Study Center in Woods Hole, Weaver specifically encouraged a seminar of scientists and government officials to investigate how the tools and methods used in weapons and aerospace systems might be adapted to solve urban problems. Similarly, in a conference presentation to the American Academy of Political and Social Science, Wood expressed his concern that many important urban policy decisions were being made without input from comprehensive data about city environments. Noting how observatories and field stations provided physical scientists with a cumulative record of the phenomena being observed, he argued that locating analogous data about urban environments, and tools for analyzing them, would help to alleviate both physical and social urban problems.[50] Some of these tools, he suggested in frequent public appearances, were the technical and technological products of military research and development.

Thus, as in the urban dispersal conversation of the prior decade, the view that much could be gained from a closer relationship between military and urban experts pervaded the highest levels of the federal government. Support, both from President Johnson and from the leadership at HUD, stimulated widespread interest in urban experimentation. HUD's backing of systems analysis and computing was not just talk. Urban observatories, urban data centers, urban information systems, an Urban Systems Engineering Demonstration Program (a program initiated by the agency under section 701b of the Housing Act of 1954 to provide federal subsidies to cities making use of systems analysis and computer technologies), and agency participation in USAC (the Urban Information Systems Inter-Agency Committee, a group of federal agencies partnering to promote improved local administration) were each components of HUD's programmatic commitment to getting these scientific decision-making tools into wider use in U.S. cities to complement national urban policy initiatives such as the Community Renewal Program.[51]

How did cybernetics' nineteenth-century meaning and twentieth-century form merge in the context of urban planning and management? The next chapter traces this story. From the early-1960s, experts from defense and aerospace found themselves recruited as advisers to management in American city governments. The partnership seemed made in heaven: better planning and management for cities, and more contracts for the defense community. Expectations were widespread that science and technology, systems analysis and computing, could be usefully applied to urban problem solving. City planning and public administration would become sciences, and Wiener's dreams of putting his work to improve human society, rather than destroy it, would be realized.

With strong federal support, theory was quickly transformed into practice. Through the 1960s and into the 1970s, systems analysis and computer simulations were brought to analyze and manage all manner of city services—from transportation planning to traffic flow to police response to housing-stock measurement. The professional literature on city planning and management (exemplified by such publications as *American City, Nation's Cities, Journal of the American Institute of Planners,* and *Public Administration Review*) is filled with discussions about these scientific innovations and municipalities' individual experiments with them. By 1968, a *Fortune* headline would proclaim: "Systems Engineering Invades the City." Professional conferences and internal city documents featured discussions about "programming" the city as if it were a computer.[52] How techniques and technologies originally designed for war became tools of choice to administer the federal Community Renewal Program in U.S. cities—and how defense intellectuals from RAND and SDC became experts on urban community development—are the focus of Chapter 3.

CHAPTER THREE

Cybernetics and Urban Renewal

In October 1964, the MIT-Harvard Joint Center for Urban Studies hosted a conference on "computer methods in the analysis of large-scale social systems." Many of the era's most distinguished communication researchers, social scientists, and computer scientists were in attendance, among them Ithiel de Sola Pool, Martin Greenberger, J. C. Licklider, Carl Kaysen, Paul Demeny, Nathan Keyfitz, and Joseph Weizenbaum. How, the participants debated, could data processing and computer simulations most productively be applied to manage and model complex systems, including cities?

As his colleagues speculated about the marvelous future that computers might offer, Weizenbaum, an MIT professor of computer science, sounded a rare note of caution. He reminded his colleagues that in the 1940s, as soon as the first atomic device was tested, many of its creators became "seized with the feeling that they had done something they wished they had not done." The capability of world destruction was not easily reversed, however, and scientific elites had to seek recourse through arms control. Weizenbaum drew an explicit analogy between computers and atomic bombs. Both were military technologies, spawned of wartime needs. In the postwar period, technology-transfer enthusiasts sought new and humane applications for computer power and atomic power. Commenting on papers and discussions that had ranged from data collection and surveillance to budget analysis and simulation, Weizenbaum expressed his fears that even the well-intentioned civilian applications of innovations in military computing might create a time bomb whose consequences could not be anticipated. "I cannot escape wondering," he mused, "whether sometime in the future we will publish a journal called the *Bulletin of the Computer Scientists,* also with a clock on the cover."[1]

In fact, a time bomb was ticking in American cities in 1964. But this time bomb, as the nation's urban crisis would become known, was not, as Weizenbaum predicted, the direct result of using computers to manage the "large-scale social systems" of cities. By the middle of the decade, when urban riots erupted across the nation, only a few municipalities had adopted military computing tools. Instead, the roots of this domestic conflict lay in failures of the nation's earlier programmatic efforts to improve the quality of city life through urban renewal.

The history of urban renewal in the United States has been told as the story of a failed federal program for urban problem solving. While in theory the Housing Act of 1949 promoted rehabilitation and housing-code enforcement, in practice renewal focused primarily on slum clearance. The "federal bulldozer," to use Martin Anderson's phrase, created as many problems as it solved by focusing attention on removing physical blight from central cities while neglecting urban social welfare in those same areas.[2] Critics argued that, by destroying the social fabric of communities that existed in slum areas, urban renewal policies in fact exacerbated problems faced by inner-city Americans. The renewal program's continued orientation toward individual redevelopment projects, as opposed to a total city plan, was also blamed; this despite the fact that comprehensive planning as a federal program dated to the 1954 Housing Act and that many cities ostensibly had crafted master plans. By the late-1950s, urban leaders, increasingly aware of the limits of past efforts, began to seek a new direction for city management.

The Community Renewal Program (CRP) typified this new direction in federal urban policy. Initiated in 1959 by the U.S. Housing and Home Finance Agency (precursor to HUD), the CRP was a large-scale attempt to turn planning from a merely physical operation to one that also focused on social factors. It was one of the agency's most visible efforts to move away from the earlier emphasis on redevelopment by demolition that had dominated urban renewal and to initiate a comprehensive and coordinated approach to renewal planning for community development. In the words of the Community Analysis Bureau (one of the governmental units responsible for community development in Los Angeles), "the safety of and happiness of the people who live in cities depends not only on the physical structures but also on their own biological and social needs and on the performance of economic and social systems as a whole."[3]

Conventional accounts of American urban development have characterized the CRP as a symbolic marker of the turn toward comprehensive physical and social planning initiatives that would be further supported by Great Society programs such as Model Cities. Yet the CRP was equally a marker of where military innovation and urban management engaged—an important example of the transfer of command-and-control-style military techniques and technologies to urban problem solving. In many cities, fears of attacks from an external enemy that drove the defensive dispersal movement did not corner the market on how the technologies and values of the cold war were applied to city planning and management. In the face of a mounting civil war in American city streets, military innovations would assume a prominent place in urban administration. Community Renewal became the first urban program to experiment with systems analysis and computer simulations.

The experiences of three distinct urban areas—Pittsburgh, New York City, and Los Angeles—offer insights into the ways that men with experience in defense analysis and military techniques and technologies made their way into urban agenda setting. As these defense intellectuals applied themselves to urban problem solving, adapting military tools for urban operations, Henry Churchill's predictions about postwar planning became a reality. "The technician," and the creation of institutional settings to support his methods of social analysis, became "an important, if not a decisive, factor in the turn of events."[4]

Pittsburgh

Pittsburgh, Pennsylvania, is the pioneering example of efforts to bring military innovations to city administration. Whereas in other cities the standard pattern of bringing information technology to local government generally began with the development of a database or automatic data processing, Pittsburgh's emphasis was on computer simulations, an approach that, elsewhere, came later. As early as 1960, the Pittsburgh Department of City Planning, after first initiating research to explore the uses of a centralized data bank, quickly shifted its focus to computer modeling. By 1962, efforts were under way to create models of city processes. By 1968, Pittsburgh was one of only two cities (the other being San Francisco) to have applied modeling in the context of its Community Renewal Program.[5]

This shift in technical focus coincided with the city's decision to hire Calvin Hamilton to direct its planning department. Hamilton came to the city from a private planning firm, but he had prior experience working for government as director of city planning in Indianapolis. Earlier, while a graduate student in regional planning at Harvard, Hamilton had worked on Project East River, the war game playing out the consequences of an atomic attack on New York City. Working with a contract research team that included two of his professors, William Wheaton and Coleman Woodbury, Hamilton had participated in some of the earliest military-sponsored planning research. It is likely that this experience influenced his ideas about directions for Pittsburgh.

When hiring Hamilton, Mayor Joseph Barr gave him explicit instructions to create a master plan for the city through 1980. Given the lack of trained staff within the planning department and the massive amount of information needed for such master planning, Hamilton contracted out much of the analytic work. Thus, while the CRP was officially a program of the Pittsburgh planning department, it proceeded with significant input from consultants at two nearby organizations. The Center for Regional and Economic Studies (CRES) at the University of Pittsburgh prepared much of the regional economic and social survey data. CONSAD Research Corporation, a think tank comprised of former RAND, Lockheed, Hughes, and Douglas Aircraft employees, relocated from Santa Monica to Pittsburgh, provided the mathematical modeling and computer simulation expertise. Researchers at the Carnegie Institute of Technology (now Carnegie Mellon) also played an advisory role.[6]

It was employees from CONSAD, who according to Garry Brewer were not originally hired by the city planning department but came on board as subcontractors to CRES, who were highly influential participants in the technology-transfer process. CONSAD was among the earliest defense-oriented think tanks to expand its client base to include "civil systems" research for both municipal and federal clients. In their role as consultants to the Pittsburgh CRP, Wilbur Steger, president of CONSAD and former head of RAND's Logistics Simulation Laboratory, and Ira ("Jack") Lowry, a CONSAD consultant who later went to work for RAND, directed urban simulation efforts at this regional think tank. Each became a pioneering example of the "technicians" that Churchill had described. As the growing use of systems analysis and computer modeling created a common language for military and urban planners and managers, both men circulated easily and frequently between the defense research community and city governments.[7]

In Pittsburgh, the comprehensive federal Community Renewal Program had multiple goals. The Department of City Planning, which organized the CRP, wanted to enumerate the city's population and its social assets, profile employment in the region, and identify the state of urban industrial, commercial, and residential land. Among the "inputs" gathered were data on housing condition, distance from business establishments, travel distance to major streets, and proximity to nonwhite areas. The data were viewed as a stocktaking to serve as the basis for identifying desired "outputs" (standard terms in PPBS) to work toward through renewal planning. The department hoped to use a multiplicity of models to simulate and forecast the potential outcomes of spending decisions for public services in different neighborhoods, for transportation planning, for city budgeting, for land use, and for social programs. By using computers, planners could constantly update the information they needed for decision making and iteratively revisit their predictions about the impact of alternative urban renewal programs as new data were compiled.[8]

Thus, any portrait of the Pittsburgh metropolitan area during the early-1960s must include, alongside accounts of urban decay and the developing suburban landscape, an appreciation of how the city was experimenting with systems analysis and computer simulations for urban renewal planning. Reports issued several times annually by the Department of City Planning under the auspices of the CRP—for example, *A Report on Poverty and Racial Discrimination* and *The Pittsburgh Urban Renewal Simulation Model*—paint a colorful picture of the range of work under way at the city agency. These documents not only provide important records of collaborative contributions to the CRP from multiple individuals and institutions but also offer a prime example of how Pittsburgh's city planners' discussions about urban "problems" shaped their discussions to fit the availability of closed-world mathematical and computer modeling tools. Consultants such as Steger and Lowry, whose analyses centered on using these innovations in military decision making for developing residential and commercial location models, predicting who would settle where and in what densities (with consequences for city administrators deciding where to provide services and infrastructure), found an enthusiastic audience in the administrators of the CRP.[9]

For example, a 1962 working study for the CRP that explained the theory behind cybernetics, systems analysis, and computer simulations suggested their potential practical applications for Pittsburgh. This document, which described the work of RAND and other contract military researchers with these tools, suggested that what planners commonly called "models" (for example, three-dimensional scale prototypes) should be understood as "simulations."[10] The 1962 study brought Pittsburgh's planners up to speed on developments in military planning and management while informing them that, in fact, what they had been doing all along was essentially the same thing. The mathematics- and computer-based simulation drawn from the defense community had many new virtues, this CRP study argued, and emphasized one in particular: the ease with which the new analytic tools could be used for prediction.

The Pittsburgh case is powerful because reports from the Department of City Planning make clear that analysts in that city, like RAND gamers, realized they were shaping their questions and problems to fit what could be modeled. In retrospect, this is a remarkable admission. Pittsburgh authors acknowledged that their models could not capture and express the complexities of reality; yet rather than characterize this as a flaw of simulation techniques, they used this finding to justify why one would want to use them. In their words, "in no case are these models photographic reproductions of reality. If they were, they would be so complicated that they would be of little, if any use."[11] The Pittsburgh team argued that, given existing technical capabilities (the computer in use at the CRP was an IBM 1401), every complex urban problem had to be defined in more narrow terms so that it could be modeled. In making this claim, the Pittsburgh team turned the original justification for the value of systems analysis and computer simulations—the tools' ability to analyze the interaction of a multiplicity of variables—on its head. Defining urban problems as requiring simplification was, in their eyes, transforming simplistic models into the ideal tool.

This philosophical stance had practical consequences. In 1964, Donald Gerwin, a CONSAD researcher, well acquainted with the CRP's multiple goals, proposed in a report for the planning department that the city construct a model based on the assumption that the CRP had a single goal: "minimizing blight." In order to use it, city planners would have to find a way to reshape their substantive agenda to fit the limitations of that particular model.

Gerwin's report for the city suggested several solutions: one was to reduce the diverse goals of community renewal to one "supergoal" (namely, reducing urban blight); another was to look for a single solution that would satisfy a minimum requirement of every goal.[12] This is an example of how the limitations of planning and management tools originally designed for a military context placed constraints on city planners and managers. Ironically, users simultaneously were concocting explanations as to why these very limitations were a good thing.

In a paper comparing how urban analysts from different backgrounds viewed the world, CRES's Charles Leven foreshadowed some of the difficulties that planners would have with the new tools. Emphasizing the political and public opinion pressures faced by planners in goal setting, he offered a contrast: that of systems analysts and economists, who focused on functional characteristics of processes, such as production, consumption, and employment. Leven cited a clash between, on the one hand, urban planners and managers, who wanted voters to be happy with the result for people, housing, and streets, and, on the other hand, the model builders, who designed systems with the goal of finding an optimum "objective" outcomes.[13]

Despite early criticisms, most model builders were optimistic that further tweaking could improve their tools. In Pittsburgh, experimentation with different models and the quest to find an acceptable fit between the evolving tools and useful practical results continued for some time. Yet these "new and improved" techniques to deal with complexity repeatedly fell short of their early promise. Even Lowry, whose *Model of Metropolis* (1965) was well-received in planning circles, admitted in his later writings that "I have deliberately avoided the question probably of most immediate interest to my readers: How well does each model work? I have avoided this question because I don't know the answers in each case and have little hope of finding them."[14] By 1973, the *Journal of the American Institute of Planners* (which earlier had reviewed and reprinted Lowry's 1965 work) would publish a "Requiem for Large-scale Models." Written by Douglass Lee, whose earlier close analysis of the Pittsburgh case was based on his own experience working there, this "requiem" disclosed that necessary data inputs for the model were never collected. Lee reported that few additional resources were invested in the project, and hence the model could neither be tested nor used.[15]

Calvin Hamilton was fired in 1964, and with his departure went many of the outside consultants who had tried to reorganize the community renewal decision-making process. Pittsburgh thus abandoned its optimism that systems analysis and computer models could actually help to improve city management. The absence of any actual uses of the system for urban policy decision making in this early effort foreshadowed the outcome of other projects. Yet the three key actors, Steger, Lowry, and Hamilton, remained undeterred, and each moved on to make his mark in other urban settings. Violent disturbances in the streets of New York City and Los Angeles would give them—and other members of the new class of urban experts—yet another motivation and rationale for adopting military tools.[16]

The Urban Crisis as National Security Crisis

In a 1946 presentation to the New York chapter of the American Institute of Planners, Tracy Augur had called for a new kind of scientific planning. "The dangers to our national well being from spreading urban blight" might not be so obvious "as those from atom bombs," he observed, yet the risks were equally grave. "Perhaps if the problems and the opportunities in courageous city planning were stated as convincingly we could look forward also to billion dollar Institutes of Research in Urbanism, to well financed Urban Development Authorities, even to Operations Crossroads devoted to the task of finding out what kinds of cities will stand up best under the conditions and the hopes of modern life."[17] Augur used an analogy between bombs and blight to call for scientific study of the benefits and dangers of urban life on the order of studies of the effects of atomic weapons.

Analogies between external threats and internal threats were found only occasionally in the urban planning community in the 1940s and 1950s.[18] But beginning in 1964, when riots erupted in several urban neighborhoods, the use of such expressions soared. That year, economist and environmentalist Barbara Ward, addressing a New York Planning Commission symposium, observed that "unless someone comes up with some jolly good solutions, the problem facing cities may become more lethal than the bomb."[19] A year later, psychologist Kenneth Clark commented that "the dark ghettos now represent a nuclear stockpile which can annihilate the very foundations of America."[20] Even observers outside the United States made a similar connection. German journalist Hans-Herbert Gotz identified a "totally new aspect of American 'Security Politics'"; in Gotz's assessment, the future of the United States did not depend on the power of its weaponry but rather on "whether in the long run human life will remain possible in big cities, whether black and white will learn to live with each other."[21] Reporting on the rash of civil disturbances, the Kerner Commission characterized an America divided by race, suggesting the nation was in the throes of a second civil war.[22] In some urban areas, this was no exaggeration—for example, in New York City, where violence escalated to the point that a city official (the director of the Office of Emergency Control Board—Civil Defense) sent a memo to the heads of all municipal agencies offering guidance on how to handle "threats of bombing and other acts of sabotage."[23] Race relations and urban problems were becoming threats to national security.

As early as 1948, Augur had observed close connections between urban defense planning and social welfare planning. Just as defense policies including dispersal might also serve social welfare needs, so, too, social welfare policies were critical for the national defense. Augur singled out a growing appetite for high-rise public housing that in his opinion made ghetto residents inviting targets for bombs. If the trend toward concentrating poverty in urban centers continued, the consequences might include a secondary threat to national security. "It is not to be expected that people who are forced to live in slums will give unquestioned allegiance to the system that keeps them there," he explained. Slums bred Communism in Augur's view, a frequent claim from public housing's detractors. (Ironically, some of these urban ghettos were at least partially a product of wartime black migration to cities, in search of work in defense industries.) Augur invoked an analogy between blight and bombs to suggest that both would be mitigated by urban dispersion, "as good a defense against internal enemies as against those whose attack is launched from outside our borders."[24]

Responses to this perceived threat of blight during the 1950s, including large-scale razing of urban "problem areas," succeeded in removing much of the blight from the daily experience of middle-class, suburban Americans. Yet the social problems accompanying those blighted areas were not so easily removed. By the 1960s, American political leaders increasingly feared exactly what Augur had predicted: eruptions of urban violence, and that widespread social ills and inequalities could be used by Communist sympathizers both abroad and at home to further fracture the nation.

Such suspicions about internal Communist and subversive activities were not imagined. Yohuru Williams has chronicled how several key figures in the emerging black power movement chose to align themselves with slogans and leaders from the Communist and Marxist world. Stokely Carmichael's visit to Cuba, coupled with scattered acts of domestic terrorism, suggested to national political leaders that the more militant wing of the civil-rights movement posed threats to domestic security. Mary Dudziak has explained how for U.S. political leaders, civil-rights reforms at home became a matter of international strategic concern.[25] So, too, calming the storm brewing in America's urban centers grew to match civil defense from the Soviet threat as a public priority. Attorney General Robert Kennedy spoke of how "something on the order of the Marshall Plan is needed" for ghetto problems.[26] Henry Maier, the mayor of Milwaukee, called for the "domestic equivalent of the military Joint Chiefs of Staff" in order to "win our war against ghetto conditions."[27] In New York City and Los Angeles, new city agencies would be mobilized for the task.

Executives and engineers from the defense community, already seeking new markets by the early-1960s, set out to transform headlined events—the urban crisis—into a domestic security challenge for which their expertise was uniquely suited. If at first they used the language of cybernetics and scientific approaches to urban problem solving, soon their justifications would expand to include the language of military attack. The fact that even before the urban crisis this community was hard at work finding ways to transfer military innovations to city administration helps to explain the appeal of reframing the urban crisis as a national security crisis, with new tactics for city planning and management as forms of civil defense. Waging war on city problems by improving urban administration became part of the military contractors' larger effort to diversify markets, to profit from the growth of federal domestic spending, and to keep their institutions in operation for the long term.

New York City

In an era when many urban leaders sought to quantify the results of their policies and programs, New York City (NYC) stands out. This is so not only because the city is unlike any other, but because urban planners' work is only a small part of a larger story. In New York City, Mayor John V. Lindsay (1966–73) gave prominent support to the principles and practices of systems analysis. Enthusiasm for military planning and management innovations at the highest level of city government led to major changes in how management was coordinated across the city. This would eventually include the creation of the New York City RAND Institute, a joint operation of the city and the think tank, in 1969.

Lindsay's administration, as well as bearing responsibility for the city's fiscal crisis in the 1970s, is widely credited with transforming New York City's management practices. Yet the city's use of military planning and management tools, including information systems, had roots in the administration of Lindsay's predecessor, Mayor Robert F. Wagner. Toward the end of Wagner's tenure in office, local government officials actively began to seek out the advice of consultants to improve continuing implementation of the city's Community Renewal Program. The CRP in New York City, under the jurisdiction of the mayor's Housing Executive Committee and the Department of City Planning, was the nation's earliest. The city received its first grant in 1960, having submitted an application to the U.S. Housing and Home Finance Agency even before the federal program had officially been established by law.[28]

A look at program documents reveals it was in New York City that the language of military strategy first made its way into urban policy discussions about eliminating urban blight. "The 'enemy' assumes many guises, including slum housing, poverty, and unequal opportunity," observed William Ballard, the chair of the city's planning commission, in the introduction to a 1965 report, explaining how "the City's existing housing and renewal programs, the extent to which they can be adapted to change, and both fiscal and human resources, represent our 'forces.'"[29] Pittsburgh's city planners also had targeted military innovations to eliminate urban blight, but their language focused on cybernetic imagery of cities as information processing systems. What differed about this document was its new rhetorical flourish. Urban blight was no longer merely a problem to be addressed with cybernetic tools; it had become an enemy to be attacked with all available resources—including military strategy. Given the city's recent episodes of rioting in several neighborhoods, this language was unsurprising. The terminology of warfare in city planning and management would become increasingly prevalent in the years ahead.

That same year, the NYC Planning Department invited Pittsburgh's CONSAD Corporation to become the lead contractor on information system development. CONSAD's work for the department took place within the framework of preparing for comprehensive planning, including, but not limited to, the CRP. Wilbur Steger headed the research team.[30] Neiland Douglas, who had worked to administer the Pittsburgh CRP with Steger while principal planner in that city's planning department, moved to become a deputy director of New York City's CRP. A September 1965 internal report produced in connection with this project described how the Comprehensive Planning Program at the NYC Department of City Planning did not yet possess adequate capacity in information gathering and analysis to do its job, making it "one of the last of the larger planning agencies in the country not to have its basic information filed in machine processing form or to have the capability to carry on planning analyses of various kinds using modern data processing techniques."[31]

The goals of comprehensive and long-range physical and social planning touted by enthusiasts such as M. C. Branch were seen as especially well suited to the adoption of computerized data processing. As an executive director of the Department of City Planning stated, "By the very nature of the comprehensive planning process, the comprehensive plan would reflect the dynamics of our continually changing urban environment. It cannot be a static, rigid, long-range ideal plan."[32] Steger optimistically proposed that the city's late start in data processing would give it the luxury of looking to the experiences of other cities for advice about data collection and model building. CONSAD researchers set out to identify the requirements for a database and to create demographic models for population and housing analyses. They also agreed to provide a series of seminars to teach the city's planners the basics of systems analysis. The goal of these efforts was eventually to open a data or information service division in the Department of City Planning modeled on the most successful such bureaus in other municipalities.

Despite his message about the potential advantages of New York City's late start, Steger was modest in his predictions about the outcomes of information system development. Technology could be adapted for use in a wide array of urban programs—for example, transportation, economic development, and housing—but only to the extent that city planners and managers first understood the complex factors behind these urban processes.[33] Moderating his claims based on his experience in Pittsburgh, Steger told his employers that he could help to make urban planning and management in the city more "scientific." Yet he simultaneously cautioned that this change in approach would not solve all the problems of urban society.

When Mayor Lindsay arrived in office in 1966, he quickly expanded plans to implement information systems across city departments and programs. In the years that followed, this would include the CRP, where contracts with Remington Rand Univac made it possible to process data from the 1960 census and then map it in graphic displays of ten-to-fifteen-block parcels.[34] But the use of such analytic tools would not be limited to city planning.

In 1965, Lyndon Johnson brought PPBS (the "brand name" version of systems analysis used at DoD) to bear on budgeting and administrative decision making across the federal government. Mayor Lindsay, who had campaigned to reduce corruption in the city by using program budgeting, followed Johnson's recommendation for federal agencies. In 1966, Lindsay decided to bring PPBS to New York City to improve budgeting and operations. He asked Henry Rowen, former assistant director at the U.S. Bureau of the Budget (and before that deputy assistant secretary of defense for international security affairs under McNamara) to serve as his budget director, but Rowen turned the job down; so Lindsay instead hired Frederick O'Reilly Hayes, who earlier had been deputy director of the Community Action Program and who had worked with Rowen during their time in the federal government.[35]

In 1966, then, systems analysis and computing came to New York City. If the city initially had lagged behind others in its synthesis of computers and urban problem solving, it quickly made up for lost time.[36] As a result, the Association of Computing Machinery (ACM) chose to hold its annual conference on the Application of Computers to the Problems of Urban Society there, and it did so from 1966 to the early-1970s, often featuring speakers from city government in New York.

Transformations in city administration under Lindsay focused first on departmental reorganizations. Thirty city departments were consolidated into ten "super agencies" to encourage more coordination and reduce duplication. Other agencies had their activities dispersed under the same rationale. For example, among the agencies dismantled by executive order was the Office of Civil Defense (established in 1950), whose activities were partitioned among existing city agencies, among them welfare, housing, and police. Barry Gottehrer, assistant to the mayor, described the period of chaos that Lindsay's changes superseded; for example, three agencies shared jurisdiction over the paving of streets; eight agencies shared partial responsibility for a winter emergency housing-repair program. Lindsay and his staff repeatedly suggested that changes to city management would soon affect the daily lives of average citizens. In his own cybernetic language, Lindsay urged "city and state governments to follow our example and improve the lines of communication and coordination" in order to "bring the good life to people in every American city."[37]

Believing that comprehensive planning could not occur without comprehensive information systems, Lindsay began to call for the computerization of more government records. In March 1969, the mayor announced that the city was "launching a computerized information system for the land and buildings in New York City" so that information could be electronically shared among city departments.[38] These technological innovations followed Lindsay's overarching strategy of centralizing power through interagency coordination and information sharing.

Lindsay hoped that New York City's approach to information system development would provide a model for other urban centers. Gottehrer's description of the mayor's plans seem almost a direct response to Henry Churchill, who had counterpoised "the roads that lead to the City of Man or the City of Eniak [sic]." Lindsay explained, "In this age in which all of our country's cities look once again to New York for solutions, these experiments to govern not by computer, but by human decision and human reason, based on sound indicators and data, are a key to the future development of the city of man."[39] The goal was clear: computerization, thoughtfully applied, would improve the quality of urban life in the nation's most concentrated urban center.

The New York City RAND Institute

Alongside his reorganizations of city government and calls for coordinated computerization, Lindsay took pains to cultivate relationships with the emerging urban management experts at RAND. RAND, the original military think tank, where systems analyses and computer simulations were contracted to military clients, would go on to play a starring role in the implementation of similar innovations in New York City. As Richard Reeves, of the *New York Times*, described it, "The city's relationship with RAND would be similar to the one RAND has had with the Air Force since World War II."[40]

Following on the work of CONSAD consultants, in 1968 Lindsay paved the way for the opening of a RAND satellite headquarters in his city by signing four contracts with the think tank. Consultants were invited to study the city's police department, fire department, housing administration, and health service administration. They were to make recommendations for streamlining operations and improving agency performance. All of these efforts taken together were predicted to reshape the city's approach to its management and operations. As Lindsay told reporter Richard Reeves, the new arrangement would bring to city departments "the kind of streamlined, modern management thinking that Robert McNamara applied in the Pentagon with such successes during the past seven years"; he called it "the most important development in the search for effectiveness in city government in many, many years."[41]

RAND's earliest foray into civilian research dated to its civil defense work in the late-1950s, some of which had included discussions of defensive dispersal. A transportation research project sponsored by the Ford Foundation offered RAND analysts an introduction to civil systems research. By the mid-1960s, RAND, like CONSAD and many other defense research institutions, had decided to turn away from its exclusive focus on military issues and apply those same rigorous quantitative methods to research on urban systems. As David Jardini has shown, this was a purposeful marketing choice in light of declining support from its major funder, the U.S. Air Force, and the fact that the Johnson administration's Great Society programs opened up new avenues for large-scale sponsored research on domestic policy issues.[42]

Before taking up his job as RAND president in 1966, Henry Rowen (as assistant director of the U.S. Bureau of the Budget) had argued that HUD, the newly established federal department for housing and urban development, needed an "urban RAND" for domestic policy and program analysis. Rowen's plan was complicated temporarily by the creation of the Urban Institute in Washington, D.C. Yet under his direction, RAND moved quickly to create its own urban institute. The choice to locate in New York City was unsurprising, given that Rowen had been Lindsay's first choice for city budget director. Thus, with financial assistance from the Ford Foundation as well as the Lindsay administration, in 1969 the New York City RAND Institute opened, a joint operation of New York City and RAND. The new unit brought a staff with significant defense-analysis experience face-to-face with the challenge of running the nation's largest city. Their studies of PPBS applied to manage social programs (for example, the NYC Human Resources Administration welfare program), rather than PPBS applied to weapons acquisition, took the systems approach in a new direction.

Peter Szanton, the first head of RAND's New York City operations, served until 1971. A "whiz kid" who had worked in the Defense Department (he was a member of the policy planning staff from 1962 to 1965) and as deputy director of the program evaluation staff at the U.S. Bureau of the Budget (from 1965 to 1967), Szanton had helped to bring PPBS to federal agencies. Ira Lowry was invited to move once again, from Santa Monica to New York City, to head RAND's New York City housing research. Central to his work there was creating a PPBS for the city's Housing and Development Administration, conducting analyses of the effects of rent control and eventually developing a voucher program.[43] Heading the health services division for its first nine months was David McGarvey, who also came to the city with extensive defense experience, including early work as part of RAND's Dispersal Team.

Although the RAND men were occasionally referred to as city planners, RAND's projects in New York City did not specifically include city planning. The topic was, however, under discussion several times. In December 1967, for example, Mayor Lindsay dispatched Donald Elliott, chair of the city's planning commission, to Santa Monica to participate in a RAND-organized workshop on urban problems.[44] In 1969, correspondence from Rowen, the RAND president, to John McCone (formerly deputy secretary of defense, undersecretary of the Air Force, chair of the Atomic Energy Commission, and CIA director), in thanking McCone for sending him the booklet "Planning for Lower Manhattan," noted that although RAND was "not as yet engaged in physical planning work in New York City . . . we plan to be in the future."[45]

The ongoing work at RAND's New York City branch represented a microcosm of changes at the parent institution, where researchers were shifting their attention from defense work to urban studies and back during this era. The personal bibliographies of several RAND consultants (for example, Robert Levine, John Kain, and Thomas Schelling) reveal how institutional changes at RAND and other defense research institutions affected the professional interests of individual researchers. This was a result of importing urban experts into defense contexts and of defense experts adapting their expertise to a new topic.

Levine, for example, a specialist on foreign political crises (he had written on Vietnam and the Phillipines) and known widely for his work on arms control, contributed to several reports about the state of the nation's cities, and he went on to become president of the New York City RAND Institute. Levine's RAND publications carry titles ranging from *A Simulation Model of Air Force Maintenance Operations* (1958) to *Two Models of the Urban Crisis* (1970). Kain, who consulted for RAND while still on active duty in the U.S. Army as a first lieutenant, wrote several reports on urban transportation and other urban topics; he went on to propose to RAND leadership that the think tank create a new program for "the study of the American Negro." Following his departure from RAND, Kain became an affiliate of the Joint Center for Urban Studies at Harvard and MIT, another institutional center for the type of urban expertise described here.[46] Schelling's work on economic strategy (pursued under the auspices of both RAND and Harvard University) moved from studies of game theory to the arms race to weapons in outer space to foreign affairs. His reports for RAND included *Models of Segregation* (1969), a mathematical study of neighborhood discrimination, and a related study, *Neighborhood Tipping* (1969), which examined the domino effect of racial change in residential neighborhoods.

The extensive time on defense studies logged by members of RAND's domestic policy research team suggests one explanation for why RAND staff increasingly came to define urban problems in national security terms and how, in so doing, they envisioned close connections between the planning and management needs of military and urban decision makers. For example, by 1970, RAND researchers were discussing the possibility of creating an "early warning system" or a "crisis anticipation system" for the mayor, a citywide evaluation system to integrate the city's numerous PPBSs, previously applied agency by agency. Rioting would be a less likely outcome if city administrators were more cognizant of public opinion, and the proposed cybernetic Mayoral Information System would take the public's temperature to assess how they perceived the performance of city agencies.[47]

CONSAD staff had laid the groundwork for city officials to draw comparisons between military and urban needs. Now, through the work of the New York City RAND Institute, city officials were learning that urban blight had become a full-fledged enemy and that urban information systems, like the information systems developed for use in Vietnam, might provide the ideal defensive weaponry. As part of these discussions, Robert Levine, at that time still in the Santa Monica office, wrote to Joel Edelman in New York about the déjà vu he experienced upon reading Edelman's memo about the Mayoral Information System. "The déjà vu refers to a project, which we have undertaken for the staff of the National Security Council, on information systems to serve national security decisonmaking" for specific geographical regions. "There is considerable common content in the Mayor's information problem and that of the NSC," Levine wrote. He proposed that RAND create a new research program, "Information Aids to the Decisionmaker," to bridge the needs of military and urban planners and managers.[48] Defining the urban crisis as an emerging national security crisis helped to make military decision-making tools seem like an obvious fit. In fact, this sort of early warning system was under development in Washington, D.C.: DEWS, the District Early Warning System—its name a play on DEWS, the U.S. military's Distant Early Warning System for incoming nuclear threats.[49]

All of this attention to urban applications of military innovations stimulated further government reorganizations. In addition to its synthesis of existing departments into ten "super agencies," New York City also created new administrative bureaus to support systems analysis and computing, including the Program Planning and Analysis Division and the Budget and Information Systems Division in the city's Bureau of the Budget. It appears, then, that the major effect of adapting military planning and management tools for planning and management in New York City, as in Pittsburgh, was not any obvious transformation in the "quality of urban life," as Lindsay had hoped; rather, the major shift was in the organization of the city administration, where program analysis techniques and data processing technologies became standard across city departments.

In its six years of operations, the New York City RAND Institute was controversial—for its politics, for its occasional secrecy, and for its expense ($75 million annually). It became the focal point for criticisms of the city's multiplicity of contracts with outside consultants for research studies of, for example, city schools, the water system, and fire department response (contractors included MIT's Urban Systems Lab, SDC, CONSAD, McKinsey, Brett and Kerr, MDC Systems, Meridian Engineering, the Vera Institute, Technomics, and Touche, Ross, Bailey and Smart). Ida Hoos, whose book *Systems Analysis in Public Policy* (1972) skewered the contracting trend, described the controversy surrounding a single RAND study, a half-million-dollar project for NYC's Housing and Development Administration. She described the "tight security" RAND imposed to suppress the study report from public scrutiny, explaining that even the chair of the city's housing committee was not given a copy. The controversy escalated when city councilors took legal action to make the report public and when, after the *New York Times* had obtained a copy and published an article on it, other members of the city government sought criminal charges against the source of the leak.[50] Hoos's version of the controversy reflected a lack of charity toward the analysts' motives. Yet such secrecy was standard operating procedure at RAND, where a history of work for military sponsors had spawned a document classification system in which some reports circulated openly while others were suppressed for internal use only. The suppressed report was such a RAND working paper, not intended for wide review.

Inside the city administration, Lindsay and his staff voiced their own complaints about outside contractors. For example, RAND spinoff SDC (the System Development Corporation) was hired to provide computer systems analysts and programmers for a data processing project at the NYC Department of Social Services (DSS). Intentionally or not, SDC contractors treated their agreement with the city much as many defense contractors treated agreements with military clients—they overran both cost and time. This created a situation where the DSS lacked documentation about the new system, meaning the city had to extend its contract with SDC so that the department could actually make use of it. In a memo to the mayor, Timothy Costello, a deputy mayor and city administrator, wrote, "SDC has the Department of Social Services over a barrel: they have not adequately documented their work to date and so the Department is essentially dependent on them at this point; no one else can do the job." A similar problem had occurred with an earlier contractor; the city's finance administration was stuck buying $400,000 in additional services from the Computer Usage Development Corporation because that company, too, failed to document its work.[51] Lacking the in-house capability for independent analysis and oversight, city administrators remained dependent on outside experts, and the New York City RAND Institute's overall contribution to cutting costs through "software type applications" was not good.[52]

Despite Lindsay's promise to reduce operating expenses, his administration was blamed for the city's subsequent fiscal meltdown. Some observers, including Frederick O'Reilly Hayes, looked back on the period to cite a dozen specific findings of the local RAND team that had been implemented or could be implemented and suggested that there could be many more. Hayes saw Lindsay's changes to city government as fairly radical, from the structure of government to the method of financing (for example, Lindsay was able to balance the city's budget for several years running).[53] RAND analysts supported this view with claims about their specific, positive contributions to city operations. These included a computer model for the city's Human Resources Administration to predict the growth of welfare caseload work. Debate about the model's predictive validity was ongoing, but its estimates were applied in preparing city agency budgets. Similarly, RAND analysts congratulated themselves on their contributions to the design of a management information system for the city's housing and urban development agency. Once in operation, the system's information on the status of building projects across the city ostensibly shifted the department's priorities "from site acquisition to development of sites already acquired."[54]

A shift in priorities did not equal a changed quality of life, however, and evidence that outside consultants had improved the day-to-day quality of life for New Yorkers was slim. More than $100 million annually had been spent on consultants whose work produced results that could be read about but not often seen. Lindsay did his best publicly to praise the work of RAND, the most prominent of the city's consultants. Yet even he recognized its limitations, conceding that city administrators were finding, like their colleagues in the federal government, that it was "far easier to establish program budgeting in the defense-like area of fire protection than in programs which have their impact, at least in part, on a man's sense of self-worth."[55] By 1970, the number and expense of outside consultants had become so extreme that the New York City Council held hearings on the subject.[56]

That the New York City RAND Institute became a target for critics of Lindsay's extravagant spending on outside consultants is ironic. Linsday had been a long-standing public critic of profligate federal defense spending (for example, in his 1970 book *The City*). His critics, however, came to use similar arguments against him, claiming that the City of New York spent far too much on outside experts, chief among them one of the U.S. Defense Department's star consulting teams, the RAND Corporation.

Abraham Beame, city comptroller under Lindsay, was one of the fiercest critics of the mayor's support for RAND and other outside consultants. Thus it came as little surprise when in 1974 Beame took the reins as mayor that the New York City RAND Institute staff would not be signing more contracts with the city. The institute closed its doors in 1975, and many of its staff moved back to Santa Monica. The New York City experiment ended with a whimper.

Los Angeles

Los Angeles, center of the U.S. aerospace industry and home to RAND and spinoffs such as SDC, was among the first users of information technology for urban data management. According to some accounts, the city first automated information gathering and analysis in 1923, when the police department adopted tabulating equipment to record and analyze crime statistics. Automation of records in the LA City Clerk's Office followed in 1946, with assistance from an IBM electric accounting machine. Adoption of automatic data processing (ADP) spread to other agencies (traffic, civil service, water and power, and police) in January 1962, when the city invested in an IBM 1401.[57]

Electronic computer programming came to the attention of the city's planning department in 1957, when the retired president of the planning commission, Llewellyn M. K. Boelter, then dean of the UCLA School of Engineering, suggested it might be worth pursuing. That year the city's planning department, under the direction of John Roberts, implemented an automated population estimate and housing inventory and began development of its City Planning Information Processing System (CPIPS). This initiated a conversation about ADP among several city agencies and led to a July 1958 contract with several consultants to explore data processing across city government.[58]

By the early-1960s, computer databases had infiltrated many city agencies, implemented in the name of increased accuracy, efficiency, coordination, and accessibility of information. The Los Angeles mayor, Samuel Yorty, like his counterpart in New York City, publicly pushed the need to coordinate information system development throughout the city. In January 1962, Yorty issued executive directive #6, creating the Administrative Services Review Committee (ASRC) to coordinate ADP centrally throughout the city. Estimates from the Mayor's Office were that rooting out duplication would save the city $230,000 annually and also reduce the need for office space. In December 1963, the mayor and the city council approved an ordinance that converted the ASRC into the Board of Administration, and the city simultaneously established a new Data Services Bureau to oversee information processing in city government beginning in January 1964. The agencies of Los Angeles city government thus began to change to reflect the realities of new technologies. The city also invited bids to look into the possibility of synthesizing the city's various uncoordinated databases into a single tool for urban management. Companies submitting proposals included SDC, the Computer Usage Company, and Advanced Information Systems.[59]

As in most cities, the Los Angeles experiments with computer simulations based on systems analysis techniques followed database development. One event in 1964 was pivotal. The City Planning Commission decided to invite Calvin Hamilton, former director of city planning in Pittsburgh, to move to Los Angeles to head the LA Department of City Planning (at the time of this decision, the commission's vice president was M. C. Branch, a professor of planning at USC who previously worked at TRW). Although Hamilton had been fired from Pittsburgh—and despite the fact that PPBS and analogous techniques had not yet documented their effectiveness in urban operations—his perceived successes were widely saluted.[60] Under his watch the department took the kinds of techniques and technologies used in Pittsburgh a step further, creating a Mathematical Model Development Program to bring systems analysis and computer simulations to aid the administration of community renewal in Los Angeles. (The city's application to HUD for a CRP grant was unsuccessful, but Hamilton viewed such programs as routes to similar ends.)

The experience of Pittsburgh suggested how practices such as urban planning and management evolve in interaction with their tools. A look at official documents from Los Angeles city departments during the 1960s reveals how—like their colleagues in Pittsburgh—planners' and managers' commitment to an image of their city as a cybernetic system in turn influenced their perception of local priorities. In "Cybernetics and Urban Analysis," a sourcebook for their course at USC's School of Public Administration, Glenn Johnson, of the LA Department of City Planning, and Leland Swanson, of USC, summarized the reasoning behind combining databases and simulations for more effective urban problem solving: "Cybernetics, the science of communications, feedback, and control, if utilized properly and adequately supported by modern electronic data processing equipment, will permit the city planner of the future to view the municipal urban environment as an integrated system rather than, as does today's city planner, to relate primarily to the objectives of the separate parts."[61] Johnson and Swanson identified two precursor steps to achieving this kind of perspective on the urban system, two steps that had yet to be taken in Los Angeles. First, data about the urban environment had to be supplied; second, information systems to synthesize and interrelate these data had to be brought into routine use in city governments.

Johnson, who was also active at the Urban and Regional Information Systems Association (he became secretary in 1970, when Wilbur Steger was president-elect), was a particularly interesting figure at the LA city planning department. His title changed several times during administrative reorganizations, reflecting the growing dominance of a systems approach to planning (his job titles between 1965 and 1970 included city planner in the research section of the planning department, division head of the planning department's Systems and Data Services Division, and planning department operations manager).[62] The creation of related jobs in city governments across the country (e.g., operations research analyst; statistical analyst; program planner) reflected the growing credibility of systems analysis in urban planning and management. Definitions of the city and its needs from men like Johnson emphasized a close fit between urban needs and cybernetic tools. This occurred even though, as in Pittsburgh, tensions between perspectives on problem solving held by old-school city planners and the new-fangled systems analysts were evident from the earliest phases of the new technologies' use.

Following Mayor Yorty's changes at the city level, Hamilton restructured the Los Angeles City Planning Department to reflect increasing attention to systems analysis and computing tools. The Systems and Data Services Division was established at the department in July 1965. The Mathematical Model and Simulation Development Section (also called the Mathematical Model Development Program) was created as part of this division, based on the assumption that "the description of a system, whether a physical system or a social system, in terms of a mathematical model, provides for increased understanding of system operations and better prediction of system performance."[63] The program's charge was to develop mathematical models to help planners to predict urban system performance and to promote rationality in planning. Its long-range goal was to accumulate a library of mathematical models related to urban planning. By 1965, the city was spending $500,000 annually on computer-related costs.[64]

Los Angeles was ideally situated during the 1960s for implementing military planning and management innovations to achieve community renewal goals. Far more than any other city, Los Angeles was home to research institutions with long-standing experience in contract military research.[65] Like RAND in the early-1960s, many of these institutions began to seek new contracts in "civil systems" analysis, civil systems that included cities. It takes only a look at the long list of characters and institutions who were consultants to the LA planning department, and the documents they produced, to see how a multiplicity of interests were well served by an image of the city as a cybernetic system.

While the work of outside contractors became a problem in New York City, it was different in Los Angeles. City council records contain numerous files approving the hiring of outside consultants, among them William Goldner, hired as an urban modeling consultant to perform a literature review of the benefits and limitations of existing urban models.[66] The planning department consulted with Britton Harris, a model developer at the University of Pennsylvania, one of a number of planners who published in such outlets as the journal *Operations Research* and whose previous work included consulting for the Ford Foundation on a plan for Metropolitan Delhi. The department also sought advice from Ira Robinson, at USC, who had directed simulation devel- ⟍ opment for community renewal planning in San Francisco, organized by Arthur D. Little. And it invited input from Ira Lowry, who by that time had moved from Pittsburgh to RAND in Santa Monica.

Alongside their work with individual contract advisers, the city administrators also developed institutional alliances with organizations including SDC and TSC (Technical Services Corporation). The planning department's main partnership for model development was with SDC, which had opened in 1957 as a spinoff of the Systems Research Laboratory/System Development Division at RAND. In its early days, SDC pursued primarily U.S. Air Force and aerospace contracts, and it is best known for its contributions to the SAGE air defense system. Yet in the face of some decline in military spending, SDC, like CONSAD, RAND, TSC, and many other defense research institutions, decided in this period to shift some of its focus toward urban affairs. Many of SDC's research staff became prominent players in the new field of urban systems research.[67]

SDC's Urban Systems Division was highly successful. It was SDC that the U.S. Department of Housing and Urban Development chose to prepare its encyclopedic report on urban and regional information systems, published in 1968.[68] Closer to home, SDC staff would play central roles as consultants to the City Planning Department in Los Angeles. All of the reports from the Los Angeles Mathematical Model Development Program series—whose subtitles included "Residential Location Models" (November 1966); "Population Projection Model Application" (January 1967); "Automated Data Inventory for the Los Angeles Region" (February 1967)—were written in collaboration with analysts from SDC.

The LA Community Analysis Bureau

Like the New York City RAND Institute, its counterpart in New York, the Los Angeles Community Analysis Bureau (CAB), also established in 1966, embodied the widespread fascination with the potential urban applications of military innovations. CAB was established as part of the Office of the Mayor, using a HUD grant of $1,360,000, augmented with a city contribution of

$680,000.[69] CAB's objective was urban problem solving in the context of renewal, to determine the causes and extent of urban blight, and to create programs that would improve both the physical and social dimensions of city life. Thus, while in most cities it was the city planning department that administered the Community Renewal Program, Los Angeles set up a separate agency to manage its version of this program. CAB performed studies of everything from the region's economic dependence on the aerospace industry to the local housing market to the concerns of area senior citizens; the agency created a land-use profile of the city and business plans for individual neighborhoods, among them South Central Los Angeles.

The creation of this new city agency was a direct response to urban unrest. Rioting in Watts in 1965, perceived to be at least partially due to failures in city administration, stimulated Mayor Yorty's efforts to address community problems with a new set of analytic tools. CAB's name alone—the analysis bureau—reflects the authority accorded to systems analysts and their approaches. Earlier, Yorty and a group of RAND staff had discussed creating a Los Angeles Technical Services Corporation, using RAND consultants to analyze a host of city problems.[70] Such an institute was never created; nevertheless, RAND's techniques and technologies rose to prominence as urban problem solving tools through the work of the Los Angeles Community Analysis Bureau. CAB hired a staff of in-house systems analysts (their job title), and city employees consulted occasionally with outside experts at organizations such as TSC.[71]

As part of its work, CAB directed the development of a number of information systems, among them the Community Program Information System (CPIS). This system was charged with using a centralized database to track what various city groups were doing with respect to community development. Computer simulations then could evaluate the impact of existing programs (and potential alternatives) on aspects of Los Angeles communities—for example, housing and neighborhood services. In doing so, CPIS and several other information systems under development for later use were predicted to become important aides to decision making.[72]

As they planned for the development of models such as an "ethnic processor which would develop probabilities of employment by job type and education," staff at CAB imagined themselves following the advice of MIT's Jay Forrester.[73] Forrester's *Urban Dynamics* (1969) had identified a common misstep in many earlier urban policies and programs—that obvious solutions sometimes exacerbated the very problems they were adopted to solve. What made Forrester's book so compelling was his claim that computer modeling of the city (as a closed system) could render visible to planners and managers the complex and counterintuitive interactions in the urban realm. To avoid the mistakes of the past, tools such as computer simulations could be adopted to anticipate these consequences before housing rehabilitation programs were fully implemented. Such tools would identify how good intentions might go awry, given that, "under certain conditions and in some areas," housing reha-bilitation "could create problems more serious than those it would treat—such as, wholesale social and economic dislocation."[74]

The Community Analysis Bureau, with its modest $2 million operating budget, may appear to be small potatoes as a historic marker. Yet its sig-nificance as an institutional center for the approach to urban decision making described throughout this book was unparalleled. At CAB, the conversation about using military innovations to solve urban problems expanded from an emphasis on cities as cybernetic systems to one that also included the notion of deterring war in the city's streets. Even more than at the New York City RAND Institute, it became clear that city administrators in Los Angeles under-stood their urban crisis as a national security threat.

Reading the documentation on Los Angeles community renewal initiatives produced by CAB shows how, in 1970, city government activities continued to be conceptualized in cybernetic terms—compared to both organisms and machines. According to one report, CAB was like a thermostat, monitoring changes in the city's temperature and making recommendations to city agen-cies about the proper course of action when it found a "difference between the desired city climate and the actual."[75] By envisioning the city agency as a ther-mostat—a paradigmatic cybernetic system—tools such as systems analysis and computer simulations (and by extension administrative units such as CAB) became ideal solutions to city problems. This cybernetic rationale im-plied a new level of accuracy in targeting city spending on urban programs—that "both the severity of urban problems and the billions of dollars expended annually on their amelioration demand that an information feedback loop monitor changes in the urban system."[76]

Other documents published by CAB in the early-1970s made explicit the increasingly prevalent notion that urban blight had become an enemy, pro-viding additional justification for the deliberate adoption of military planning and management tools as a response. City officials were confident that "the plight of the city surviving the threat of urban blight in its broadest definition suggested a common thread with the philosophy of weapon system development of the Department of Defense, which starts with the mission analysis comprised of threat assessments and requirements to counter those threats."[77] Using this analogy, they suggested that the work of CAB and the Department of City Planning was to "process threats" by surveying city neighborhoods and developing "action programs" as a response.[78]

Rationalizing the fit between military innovations and city needs, CAB staff drew analogies between the threat of an external military attack and the threat of an internal urban crisis. Just as defense information systems such as SAGE offered real-time analysis of incoming information, CAB analysts suggested how an urban analog based on records of neighborhood deterioration might create an "early warning system" for possible areas of riot activity and propose the kinds of social programs and infrastructure plans that would thwart the threats. The parallels with respect to threat assessment and response were particularly evident in a 1970 report that produced a pair of flow charts—Weapon System Development Flow and Urban System Development Flow (fig. 3). Both began with a "threat assessment" and moved toward action steps. Like their colleagues at the New York City RAND Institute, CAB staff envisioned a combination of urban programs and information systems creating a new method of crisis anticipation for city decision makers. The CAB director, Robert Joyce, saw his agency's work as reflecting the history of war as an engine of innovation. "Until the Watts riots erupted in 1965," he observed, "no efforts of consequence had been made to define the underlying problems in that community." But immediately afterwards, "a plethora of programs were initiated on a crash-type all-out approach."[79]

The culmination of CAB's research work in Los Angeles was the release of a four-volume report, *State of the City*, beginning in 1970. By that time, information systems in use in the city included an Automated Planning and Operations File and a Los Angeles Municipal Information System. Other systems were still being developed; for example, a Los Angeles Urban Information System and a Comprehensive Urban Simulator—to "support analysis of blight and building obsolescence in the city" and to use simulations to decide whether the secondary effects of proposed housing rehabilitation would in fact create problems more serious than those it would treat.[80] The report outlined the failure of urban renewal, described the "threat to urban life" that remained (the title of volume 2 was *A Strategy for City Survival*), and called for a

WEAPON SYSTEM DEVELOPMENT FLOW

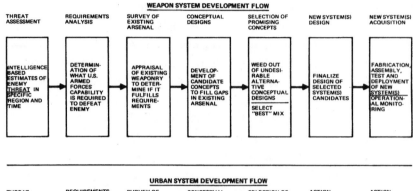

URBAN SYSTEM DEVELOPMENT FLOW

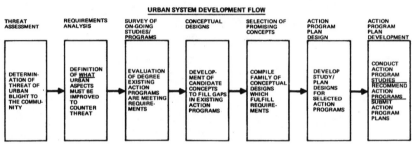

Fig. 3. Chart showing how city planners and managers in Los Angeles directly drew upon the expertise of military planners and managers in their efforts to attack urban blight. The chart, part of a study prepared by the Los Angeles CAB, compares urban systems with weapon systems. From *Design Requirements for the Data and Systems Support Essential to an Urban Blight Systems Analysis* (Los Angeles: Community Analysis Bureau, 1970). Reprinted with permission of Los Angeles City Archives.

comprehensive survey and analysis of existing programs and future possibilities. Volume 4, in particular, was an homage to systems analysis, praising the "weapons system analogue" for assisting in the "problem definition and program formulation" process and arguing for increased quantitative evaluation in cities as a feedback mechanism for urban managers.[81]

CAB leadership acknowledged public criticisms of their New York City colleagues and the increasingly widespread notion that "the system man is often referred to as a man with a solution looking for a problem."[82] They also offered the caveat that models from military and aerospace contexts, and even other management contexts, were not easily transferred to urban settings—meaning that any new models would require multiple revisions.[83] They acknowledged that "the only way to test completely the adequacy of the analysis and the new concepts" drawn from military sources was to engage in war and that, "because the system is supposedly designed to prevent that eventuality," no one could "truly test its effectiveness."[84]

Yet such criticisms did not deter the bureau's call for further applications. The *State of the City* report discussed tools for future implementation, including program management, information systems, cost-benefit analysis, PPB, performance specifications, statistical analysis, program impact and analysis, and simulations. Proposed projects for future study included a systems analysis to analyze the impact of the city's new convention center and to predict the effects that the Martin Luther King Hospital in Watts (still under construction in 1970) would have on city communities.

So whereas by the late-1960s Pittsburgh had tired of mathematical modeling, and New Yorkers were becoming uncomfortable with the levels of taxpayer money given to outside consultants, Los Angeles offered a sharp contrast. Through the mid-1970s, city officials pressed for additional resources. Much as military planners working on the SAGE air defense system never had the opportunity to test their innovation in an actual war, city planners in Los Angeles, like those in New York City and Pittsburgh, gained more status from the planning process than from applications of their analytic tools in the field. In LA, report after report continued to emphasize future promise, and despite few visible positive results of the investments in new methods and tools for analysis, the technology-transfer momentum endured.

Evaluating the Results

The experiences of Pittsburgh, New York City, and Los Angeles demonstrate, on a small scale, the variety of planning and management techniques and technologies based on cybernetics and computer simulations that were imported for use in cities. While some areas chose a city-focused information system to improve budgeting and the delivery of municipal services, others participated in regional transportation planning projects. While some made use of PPBS off the shelf, others developed specific software such as the Basic New Community Simulator (NUCOMS), a computer system designed to evaluate economic and financial potentials of new community development.[85] While some urban information systems aimed to collect information for a citywide database to analyze urban blight, others used mathematical modeling toward monitoring specific community renewal programs or forecasting the potential outcomes of different policy decisions.

The history of postwar interactions among planners, city managers, and defense experts helped to make the adoption of systems analysis and computing seem more natural than artificial. Yet technology transfer, or even the basic adoption of any technology, is a complicated negotiation. Revisiting community renewal and other efforts to improve urban operations in light of this understanding of technology transfer illustrates how urban planners and managers redefined urban needs to fit the requirements of their military tools.

Over time, a fundamental dilemma emerged across cities. The techniques and technologies widely used for urban problem solving never transcended their military roots. In most cases, the adoption of these tools did not show obvious gains for urban residents, especially in the area of community development. Tools that had proved reasonably successful in military contexts became the objects of far greater scrutiny in their urban applications. The emphasis on evaluations at the federal level, instantiated by a multiplicity of federal Offices of Program Evaluation (and the development of the field of "evaluation research") has left a historical legacy of evaluation literature that can be mined for general conclusions about these urban experiments.[86]

Sociologists who study public problems often seek to identify how a particular construction of a problem "works" for a prominent group of claims makers. The examples presented here document how problems and potential solutions in three major U.S. cities were redefined in terms that would facilitate the adoption of a particular set of tools. This redefinition process served the needs of multiple constituencies, from urban managers seeking to look like they were doing something about urban problems to defense research organizations seeking new contracts. Yet when it came time to demonstrate not just rationalizations but concrete results, the techniques and technologies that had come to be seen as direct responses to the deficiencies of older urban planning and management practices did not deliver on these promises. The increasingly wide use of systems analysis and computer simulations in cities became characterized as a "revolution," but evidence from the nation's cities testified that a revolution in the quality of urban life was not forthcoming.[87] The most strident critics suggested that the rhetoric of problem solving masked the reality of other interests. In the case of approaches to the urban crisis, Kenneth Clark and Jeannette Hopkins put their contempt bluntly: "The rhetoric of involvement of the poor, combined with studied control of the poor, has led to ineffectiveness, to a substitution of rhetoric for basic social change."[88]

Cities had adopted technologies under the guise of widening the scope of urban problem solving. And mayors had learned to speak the language of "problem orientation." Yet the innovations they chose, while specifically adopted as a means to broaden urban problem solving under conditions of uncertainty—to choose the decision path with highest probability of best results in a world of contingencies—nevertheless were unable to capture the full range of complexities in the dynamic urban realm. The definition of urban problems that these tools could accommodate remained narrow. By 1970, Philadelphia city planner Abe Gottlieb chastised his colleagues: "Any seeing sentient person," he said, had to acknowledge that "smoothly operating data flows within and among city departments," alongside improved object-oriented inventories of buildings, land parcels, and municipal service locations, were not remedying the nation's complex urban problems.[89] Successes increasing the efficiency of firefighter dispatch and police dispatch—among the few visible, positive outcomes of adopting the new tools—were not accompanied by remedies for more complex problems of urban social welfare. Computers would not serve these needs until they could be equipped to generate solutions beyond the efficiency of municipal operations.

Even some claims of success, when examined closely, reveal that the outcomes could not definitively be linked to the new techniques and technologies. An example of this is found in Dayton, a city with long-standing ties to the U.S. Air Force (home to Wright-Patterson Air Force base) and also one of HUD's Model Cities. From the late-1960s, the city expressed interest in developing a municipal information system, in particular to use PPBS to improve public financing. This project, modest in its scope, became a reality thanks to financial assistance from HUD's information system program and a partnership with the University of Dayton and Westinghouse's Civil Systems Division. At a conference in 1970, Nicholas Meiszer, assistant to the city manager, reported highly positive results from these new tools—increased coordination between local and federal government, stronger ties between the city and its university community, and time saved. Yet careful analyses of those findings reveals that only this third outcome—that data processing estimated to take twenty-six "man years" could be compressed into two—could explicitly be linked to the new techniques and technologies.[90]

Several factors stand out to account for the lack of positive results. First, while it was widely argued that the new techniques improved on older urban renewal programs because of their "scientific" nature and because they systematically could combine physical and social planning, a lack of relevant data (particularly on social welfare matters such as quality of life and quality of health) stunted the uses of decision-making tools that were heavily reliant on quantification. Systems analysis and computer simulations could not live up to the promise of measuring the "outputs" of social policies and programs when social data inputs were missing, or not easily quantified. Technologies adopted to facilitate social planning were thus handicapped from doing so.[91] This argument explains the strong pressure on planners to stick to the old-style physical planning that had defined the earliest urban renewal efforts. Despite its good intentions, then, the cybernetic turn in urban analysis failed to displace the dominance of physical planning in renewal programs.

A second explanation for the tools' ineffectiveness, the challenge of goal setting, followed directly from the difficulties of quantification. Goal setting was a critical step for making effective use of the new methods. Some of the difficulties city administrators encountered stemmed from their lack of deep understanding of the tools; often urban planners and managers began using them without going through a process of a priori goal setting. These blind uses proved ineffective when the assumptions of the tools—most often that saving money was the best outcome—drove decision analyses.[92] Savvy administrators counseled colleagues to program their own goals into the software, yet even when urban planners and managers took active roles in goal setting, all did not run smoothly. Social goals such as eradicating blight and improving the quality of urban life simply were not quantifiable in the precise sense that military objectives could be. Mathematical models could not capture the complexity of goals in urban settings.

An example from Pittsburgh illustrates the difficulties of the goal-setting process. Even though Pittsburgh's planners and model developers were aware of the need to set community renewal goals, the limitations of their tools meant that multiple goals had to be reduced to one. As Charles Leven observed in his analysis of Pittsburgh, planners and model builders disagreed as to what this "supergoal" should be, making effective model building nearly impossible.[93] Leven's clash-of-cultures argument offered an opposing view to claims that systems analysis and computer simulations simply were the latest generation of models for a profession long used to models. And it foreshadowed a power struggle that played out in several cities, a struggle between more traditional urban professionals and the new systems and computing experts.[94] Even Frederick O'Reilly Hayes, an enthusiast, conceded that in New

York City the information system development process more often than not followed the technical needs of systems analysts, rather than the goals set by city managers.[95]

This power struggle suggests a third explanation for the disappointing outcomes: the adoption of systems analysis and computing accompanied the urban professions' struggle to reassert themselves soon after the massive and very public failure of urban renewal; yet just as these new tools were offering hope for the profession, simultaneously they began to undermine its authority. Technocratic systems and defense analysts, experts in the new tools, became a new class of urban experts, publishing in leading city planning and management journals, joining the staffs of city planning departments, and consulting to mayors' offices across the nation. Whereas the defensive dispersal movement had helped city planners to rise to prominence on the heels of national security concerns, now national security experts were displacing city planners.

Some commentators interpreted these changes in positive terms. For example, Leland Swanson and Glenn Johnson observed that new technologies altered city planners' social position. As they became part of a larger systematic endeavor—the "highly complex probabilistic system" of municipal government—planners became technical advisers to urban management.[96] Anthony Downs, an expert on real estate and urban affairs at the Brookings Institution, saw similar benefits. At the 1967 conference of the American Society of Planning Officials, Downs argued that planning was in the middle of a revolution, which he characterized as moving power from architects and designers toward a focus on many nonphysical elements of urban and regional management. He described the new approach as more dynamic and better equipped to integrate state- and federal-level planning activities into a coordinated approach. In this problem-oriented approach, planners' power was reduced for the greater good of the system. Both were benign interpretations of the power shifts made possible by new technologies.

More critical interpretations of the power shift saw city leaders—mayors and planning department heads—handed a tool to consolidate their power under the guise of "objective" best results. John Kolesaar described how PPB systems merely served as "window dressing" for savvy users who knew the answers before they asked any questions.[97] This was perceived to have occurred at the federal level, where, according to Downs, Defense Secretary McNamara used PPBS to assert his authority over each branch of the armed services, and

Downs believed that John Gardner was using similar tactics at the U.S. Department of Health, Education, and Welfare.[98] The most strident critics (for example, Ida Hoos) argued that systems analysis, whose successes in military contexts begat enthusiasm for its domestic uses, in fact had failed in its wartime applications. In her words, "like the military successes in Vietnam, the fictitious accomplishments of PPBS are still put forward as though actual."[99]

Others observers offered less rosy analyses of the changing professional dynamics in urban planning and management. The most extreme depictions suggested cities were helping to achieve the darkest consequences of cybernetics—government by computer. This was not a case of human-like computers taking over the world, but rather humans coming to think more like their machines, limited in their beliefs of what it might be possible to accomplish in cities. Eugene Nickerson put it bluntly: computers understood the world in terms of "dwelling units," but people saw it in terms of "homes." The distinction was "more than just semantics," he observed, for "as the failure of so many urban renewal programs has shown, what's good for the dwelling unit isn't always good for the home."[100] Such critics warned that urban professionals already had come to think more like their computer tools.

This notion of the clash of cultures between computers and humans points to a final explanation for the limited successes of technology transfer: fundamental distinctions between the organizational culture of military versus urban decision making. Military organizations have a hierarchy of command and control; hence, orders from the top are implemented with only rare disagreements or feedback. By contrast, authority in complex city governments is far more dispersed; city agencies rarely march in lockstep. Many defense intellectuals (as well as organizations such as URISA, at least in its early years) assumed that a master plan, carried out with more centralized coordination, could improve the plight of American cities. But as urban administrators recognized early on, cities are far more complex organizations than the military, and master planning can be a controversial exercise. While most Americans could agree that the United States was worth defending, the goals of specific social programs, by contrast, were not so easily universally approved. As William Ross, deputy undersecretary for policy analysis and program evaluation at HUD, noted, "the protective blanket of national security is not applicable in our civilian world at the federal level—and certainly not in the goldfish-bowl situation at local levels."[101] At a time when calls for citizen participation, transparency, and more accountability to the public eye in urban management were growing, tools created for a more closed military setting—even effective tools—clashed with such calls for openness in goal setting, as the case of the New York City RAND Institute demonstrated.

Thus, when General Bernard Schriever, USAF (retd.) wrote "Rebuilding Our Cities for People" in *Air Force and Space Digest* (August 1968), recommending that cities establish a single centralized management authority, he ignored how that recommendation for command and control sat at odds with requirements for citizen participation in the CRP and other urban programs.[102] Military information systems such as the Hamlet Evaluation System, created in the context of war, did not have to generate recommendations to appease a democratic public. Yet as a 1974 evaluation prepared for HUD by researchers at Howard University made explicit, one of the two main barriers to the implementation of urban information systems was the simple fact that some of the proposed solutions based on models were "politically unacceptable . . . for example, proposing an incinerator or landfill in an area that would not accept it, or proposing land uses that land developers opposed."[103] Peter Szanton had recognized similar difficulties in his work with New York City. For example, the suggestion made by RAND staff that the city eliminate certain fire stations in the name of economy was highly unpopular.[104] The implementation of military techniques and technologies ran up against barriers (e.g., cooperation among city departments or cooperation among citizens and local government) that engineering could not fix. While military commanders such as Schriever could compel their subordinates to fix these problems, that was not possible in city government.

Understanding the fit between military innovations and urban needs as an intellectual construction reveals how the fit was coming undone as soon as the techniques and technologies were first applied. There was overwhelming evidence—from a lack of positive outcome in individual cities to negative evaluations of PPBS in federal agencies and state governments—that technology transfer from military decision making to civilian arenas had not been a success, and in some cases the emphasis on technology had been a diversion from the most urgent urban problems at hand.[105] Early critics were muffled by those who believed technological improvement would lead to better outcomes. Even many of the critics expressed a fundamental optimism, suggesting that there were possibilities for improving either the tools or the process of implementation.

For example, the suggestion that software imported from military applications reinforced hierarchy in decision making was accompanied by the proposal that it be modified—to incorporate a more democratic decision-making process. According to this view, municipalities had yet to reconfigure data systems to suit their purpose, and old uses were driving new technologies where imagination could do much good. Automation expert John Diebold was optimistic when he compared his era's computer experts to "the early auto-makers who at first thought they were making 'horseless carriages'" but later moved far beyond that idea. Participatory models such as the Dayton Neighborhood Achievement Model were designed in this spirit.

By the mid-1970s, however, the critical view had become more widely accepted, as the failure of the Vietnam War helped to end the era of confidence in these techniques. Paul Starr has called 1973 the twilight of simulation, and, indeed, around that time some prominent institutions shuttered their systems analysis and computer simulation shops.[106] The ACM suspended its annual conference after the 1973 meeting. MIT closed its Urban Systems Lab in 1974. RAND dismantled its New York branch in 1975. The federal interagency committee, USAC, held its last meeting in 1977.[107]

Yet the infrastructure of institutional support for experimental approaches to city governance that accompanied the cybernetic turn in urban analysis would have lasting effects. Think tanks and government laboratories such as RAND, SDC, and Oak Ridge continued to operate "urban systems divisions" and "urban and regional studies departments." Cities continued to hire outside consultants to analyze and improve their management and operations.[108] Schools such as Harvard and Carnegie Mellon, educating the next generation of policy analysts, continued to teach "decision sciences" and "public management science." Federal agencies maintained Offices of Program Evaluation and Analysis. Prominent national and state organizations (for example, the American Society of Planning Officials, the International City Managers' Association, the Council of State Governments, the Urban and Regional Information Systems Association) continued to hold conferences and issue reports on how to implement effective municipal information systems. A journal that began publishing only in 1975, *Computers and Urban Society,* with a cast of familiar characters at the helm, renamed itself *Urban Systems,* and then *Computers, Environment and Urban Systems* (it still publishes today).

Thus, like the defensive dispersal movement before it, the lasting significance of this episode in the history of American cities cannot be measured by the direct effects and successes or failures of systems analysis and computing in the Community Renewal Program. The apparent failure of technology transfer in this endeavor does not negate its importance in the story of American urban development. True, innovations based on military research and development did not lead to obvious improvements in the quality of urban life. Yet the long process of trying to adapt military innovations for urban use forged stronger alliances between defense experts and city administrators. Individual cities' efforts linked them into an emerging network of individuals and institutions—think tanks, federal programs, and other cities—supporting the new marriage of technique and technology with the "science" of urban public administration. Together with organizational changes within departments of city planning and across city governments, including the creation of new city bureaus, these were among the lasting effects of transferring military innovations to American cities. Paul Dickson suggested there were more intangible consequences, too. Meditating on the troubled history of RAND's New York office the year it closed, he observed: the institute's "most profound influence" was "on the nation's way of looking at things" and "demonstrating that analytic techniques have a role to play in dealing with street-level problems."[109]

Most important, the cybernetic turn in urban analysis became an episode of continuing significance as early experiments facilitated the rise to prominence of defense intellectuals who worked hard to become experts in urban problem solving. These were the technicians Henry Churchill had described. Their faith in technology, and the faith that others had in their abilities, persisted long after specific failed experiments. Thus, while their first efforts with systems analysis and computer simulations were not a grand success, these technicians did not abandon hope. In the years immediately following, defense intellectuals at several think tanks and their close cousins in the aerospace community would find other opportunities to participate in urban problem solving, using a different set of technical and technological tools.

Part II / Cities in the Space Age

CHAPTER FOUR

Urban Intelligence Gathering

In 1968 remarks to a forum on "systems analysis and social change," U.S. Vice President Hubert Humphrey offered his analysis of military technology transfer to date. The nation's military-industrial complex had developed a robust variety of managerial innovations, he observed, but there was still much that the aerospace community could do. He presented the assembled audience—members of the American Institute of Aeronautics and Astronautics and the Operations Research Society of America—with a challenge: "I want every one of you to become more involved in solving our problems here on Earth," he declared, citing the troubles plaguing the nation's cities as an especially urgent priority.[1] Humphrey, chair of the National Aeronautics and Space Council and a former Minneapolis mayor, suggested that if aerospace executives and engineers followed his call to find new markets—a sensible strategy for any industry—they might lead the way toward better governance in America's increasingly ungovernable cities.

Humphrey was not the first to suggest the aerospace industry's ideas and innovations could offer assistance to the nation's city administrators. Yet his remarks offered a new motivation for their intervention in urban affairs. "The techniques that are going to put a man on the Moon are going to be exactly the techniques that we are going to need to clean up our cities," he explained in another presentation that same year, pointing to the "systems analysis approach" as an especially promising strategy.[2] Humphrey's proposal was a direct response to criticisms from mayors such as Detroit's Jerome Cavanaugh, who charged that excessive spending on the Vietnam War and the Apollo Program was diverting investments from U.S. cities, even as NASA's budget was curtailed. Humphrey painted a picture of cities improved, rather than depleted, through an expanded aerospace industry.

At the same time that their colleagues in think tanks worked to bring cybernetics and computing to American urban operations, the aerospace community proposed that its own brand of innovations would offer an ideal complement. Systems analysis was not, as Humphrey predicted, their central focus. Instead, innovations in nonphotographic reconnaissance technology and image-interpretation techniques to survey territory from air and space, as well as the "space age management" techniques that were said to have made possible these innovations' rapid development, became the aerospace community's chief proposed exports to urban markets. Like their colleagues in defense research and development, this breed of technology-transfer enthusiast found ways to define city problems to parallel problems encountered in the nation's space program. Improving the comprehensive planning process in an era of cities' limited financial resources was the overarching stated goal of their initiatives. The unspoken agenda was repairing the aerospace community's public image in the face of critics such as Cavanaugh in order to insure the industry's survival.

U.S. defense and aerospace communities historically have been tightly coupled, yet one significant difference separated their experience in urban operations. For the defense intellectuals who worked to export military innovations to Community Renewal and related programs, declaring a "war" on urban problems became a controversial exercise. Activists charged a war already was under way in America's inner cities—a war against, rather than for the benefit of, urban populations. The community of aerospace experts who followed Humphrey's advice faced a far less daunting climate for public relations. For the activities of the military space program remained wholly classified, leaving the aerospace community with a civilian agency to lean on. With NASA as their figurehead, aerospace executives and engineers eschewed the public rhetoric of war in favor of the language of scientific planning and management. Yet if their linguistic ballet painted one picture of industry efforts to promote new approaches to urban data gathering and analysis, a more accurate representation remained concealed. Behind the rhetoric of science lay the reality that the innovations that aerospace executives and engineers endeavored to transfer to urban operations—like NASA and indeed much of the aerospace industry—were historical products of America's cold war concerns about national security.

A Brief History of Aerial Reconnaissance

Military leaders have long depended on remote sensing techniques to gather information about enemy territory. The earliest recorded uses of aerial surveillance in the West date back to the French Revolution. In April 1794, a corps of *aerostiers* participated in the war effort from high above battlefields in balloons, with some of these air warriors continuing to serve Napoleon in the years that followed. Photography, invented around 1820, made it possible to keep records of the view from the air. In America, the first aerial photos were taken over Boston in 1860. Under President Lincoln, the U.S. Army created an Army Balloon Corps, although these balloonists played only a small role in the Civil War.

The history of the U.S. aircraft industry, like its successor the aerospace industry, has been closely tied to military needs.[3] The earliest known images gathered from an airplane were L. P. Bonvillain's motion pictures of Italy, taken from a civilian aircraft flown by Wilbur Wright in April 1909, but the practice of aerial photography quickly became identified with military strategy. Two years later, in 1911, when the U.S. Army Signal Corps created a flight school in Maryland, it put aerial still photography on the syllabus. World War I saw the first major use of military reconnaissance flights using aerial photographs. Reconnaissance, previously the job of cavalry, became so popular by air that "air units were even given cavalry names: squadrons."[4] And the new job title of "air planner" (later replaced by "strike planner") arose to refer to the group of military personnel whose task was to plan campaigns from the air.

It was commonly said during the 1910s and even into the early-1920s that aerial photographs, due to their distorted representations of objects, bore striking resemblance to abstract paintings.[5] This was a serious impediment to using aerial imagery as a mapmaking technology or in any other context of strategic planning. Following World War I, aerial photographers—including those in the U.S. Army Air Service—set about to make major strides in technological development and interpretation techniques to reduce distortion. Their interwar innovations in technology and the rise of increasingly expert image interpreters would serve military efforts well when World War II arrived.

The uses of aerial reconnaissance in World War II were both offensive and defensive. During the war, aerial photography expanded its range of uses to play a major role—in mapmaking, target identification, and assessment of enemy capabilities.[6] Declassified images from the U.S. Defense Intelligence Agency and other branches of the military services at the National Archives reveal several million images of overseas territories, among them numerous aerial views of foreign cities. Strategic decision making at the climax of the conflict—the bombing of Hiroshima and Nagasaki—depended upon aerial surveys, and the federal government's Strategic Bombing Survey (which created an Urban Areas Division sometime between 1946 and 1947 and included Paul Nitze, Harry Bowman, and John Kenneth Galbraith on its staff) continued to publish aerial studies in the years that followed.

As part of wartime mobilization efforts, private companies and university scholars were recruited to assist in aspects of combat preparation that involved aerial photography, its interpretation, and its applications to mapmaking. Thus, the civilian aerial survey industry, itself a product of expertise developed in World War I, found many opportunities to participate in World War II. For example, the U.S. War Department and the Army Corps of Engineers invited the Los Angeles branch of Fairchild Aerial Surveys to collaborate in mapping the Southern California region in 1942. Other companies such as East Lansing's Abrams Aerial Survey Corporation served as course directors for aerial mapping and airphoto instruction of the U.S. Army, Navy, and Marine Corps. Kodak suspended production of its civilian Kodak 35 and exclusively sold it to the army for use in the U.S. Signal Corps' photographic field kit. Similarly, for many academic geographers World War II created new opportunities for jobs in government service and public funding for their research. Early in the conflict, military planners found themselves lacking reliable maps and so they called for better accounts of terrain. The large-scale mapping effort that enlisted the services of private aerial survey companies also took numerous university geographers to Washington, D.C., to work in agencies such as the U.S. Geological Survey and the Office of Strategic Services (precursor to the CIA), as well as overseas. Aerial photography was a key tool in these mapping initiatives, and records in the National Archives reveal the comprehensiveness of efforts to catalog foreign territory using aerial views. Thus, aerial photography had an early and important place in the military-industrial-academic complex that so many U.S. historians have described.

Relationships developed in wartime would be of long duration. Following the war, private survey companies continued to seek relations with military and intelligence agencies, for example by sponsoring photo-intelligence units as part of military reserves programs. University scholars whose appetite for federal funding had been whetted found federal support from the Office of Naval Research (ONR) and NASA, stimulating the creation of academic programs in photointerpretation (which in turn became programs in remote sensing). Thus, while in 1950 only eleven U.S. geography departments taught aerial photographic interpretation, by 1973, ninety-seven schools were teaching the subject.[7] Looking back on the period, Kirk Stone, a geographer in the Office of Strategic Services during World War II and later a professor of geography at the University of Wisconsin and the University of Georgia, observed: "World War II was the best thing that has happened to geography since the birth of Strabo."[8]

Wartime exigencies did more than stimulate increased contact among industry, academic geographers, and military and intelligence agencies: they also led to further refinements in techniques and technologies for photointerpretation, surveying, and mapping. The U.S. armed forces created some of the earliest photointerpretation manuals during the war, and the increasing professionalization of photointerpretation would expand the range of details that could be read from reconnaissance imagery.

Chapter 2 described how in the climate of the early-1960s, defense and systems analysts at institutions such as RAND began to turn their attention to the social sciences. Studies of political systems conducted in an international context as military and intelligence contract research (e.g., in the Vietnam conflict) reflected an expanding vision of U.S. security strategy, one that embraced an interest in the socioeconomic and demographic dimensions of target populations. The history of airphoto interpretation parallels this trend in national security studies. From the late-1940s, the uses of aerial surveys for military and intelligence agency sponsors began to move beyond inventories of physical infrastructure of urban targets to attempt to intuit socioeconomic information about urban populations. As one proponent explained, photointerpretation data did not directly provide "social data," yet they were "pertinent to social research needs in so far as such 'physical data' have meaningful sociological correlates."[9]

M. C. Branch, one of the earliest authors on systems analysis and computer simulation in city planning, was even earlier a proponent of developing photointerpretation techniques to discern socioeconomic information from aerial images. His training in photointerpretation with the U.S. Navy, combined with his wartime experience as a research assistant on Roosevelt's National Resources Planning Board, led to his book *Aerial Photography in Urban Planning and Research* (1948) soon after the war. Branch stressed how advancing photointerpretation techniques made it possible to extract a wealth of information from aerial surveys of urban physical infrastructure. For example, population densities could be estimated based on the type of residential structure. So, too, "neighborhood situation and character" as viewed from the air could be "sociologically revealing," with land values discernable based on physical features such as type and number of buildings and landscape."[10] Branch's emphasis on how socioeconomic data might be extracted from such surveys signaled a new direction in photointerpretation.

One of the earliest studies devoted to aerial photography as a tool for remote collection of demographic and socioeconomic information was a 1953 report by two U.S. Air Force officers, Maj. Norman Green and First Lt. Robert Monier. Given a "restricted" classification, the report was prepared for the USAF's Human Resources Research Institute as part of POP KEY (Population Keys to Urban Areas), a program initiated in 1950. (Green and Monier also supervised the contract research on urban dispersal performed by Philip Hauser, Otis Dudley Duncan, and Beverley Duncan, of the University of Chicago.) The authors explained their motivation: In wartime, certain areas were not physically accessible. This made target populations impossible to contact. Thus, an indirect survey method was needed to assess the situation. Green and Monier explored and evaluated the reliability of aerial reconnaissance as a method for collecting and interpreting social data, with "ramifications upon target selection, target analysis, and human factor description assessment . . . valuable to Air Force operations in strategic bombing, Psychological Warfare, and in intelligence long range planning."[11] Like so many studies of aerial reconnaissance techniques and technologies of this period, they began by testing black-and-white aerial photography over U.S. territory (in this case, Birmingham, Alabama), comparing aerial information based on air force imagery from 1947 at a scale of 1:7000, with housing data from population estimates collected on the ground. The authors concluded that this method needed work before it would become reliable, but they were hopeful about future uses of more sophisticated photointerpretation techniques combined with new reconnaissance technology.

At the same time that military photointerpretation was expanding its possible uses, efforts were under way to develop nonphotographic sensor technologies to expand the uses of aerial surveillance in military operations. During the war, new sensor technologies (including radar and infrared detection) had made it possible for surveillance aircraft to record information beyond what could be perceived by traditional, visible-light photographs. In the postwar period, an expanding military reconnaissance program stimulated further development of still more nonphotographic techniques such as SLAR (side-looking airborne radar) and multispectral scanning. These innovations in sensor technologies would help to bring reconnaissance into the space age.[12]

Cold War Satellite Reconnaissance

Space flight has been a long-time staple of mythology and science fiction, dating back centuries. But as science fact in an American context, its roots are intimately bound up with the military-industrial-academic complex described throughout this book. The earliest known (i.e., declassified) photos of Earth from space were taken from cameras attached to rockets in 1946. The idea of shifting military and intelligence reconnaissance operations away from manned aircraft and toward unmanned space vehicles was first raised in a report that same year by staff at Douglas Aircraft. These Douglas researchers, pursuing contract work for the U.S. Army Air Force (soon to be the Air Force) would become the earliest employees of the RAND Corporation.

A U.S. military space program dominated by the air force and supported by industrial allies such as Douglas predates the creation of NASA, and RAND's participation provided critical research support and analysis. Thus, as U.S. military and intelligence reconnaissance programs continued operations using aircraft surveillance, during the late 1940s and into the 1950s momentum was building to transfer at least some operations to space-based platforms. Satellites were predicted to be an especially useful component of the armed forces' strategic arsenal because compared with planes—even high-altitude aircraft such as the U-2—they would be harder to knock out of the sky.

The initial years of the American space program were characterized by territorial battles and jockeying for position among each of the armed services as well as the growing intelligence community. In fact, in the years immediately following World War II, the U.S. Army, Navy, and Air Force each rushed to create a military space program, and the CIA, too, wanted a piece of the pie. With the highly publicized launch of Sputnik, the Soviet satellite, in 1957, the U.S. government reconsidered the direction of its developing military space program. The government decided that in addition to maintaining secret programs to conduct highly classified space research for military and intelligence work, the nation should create a civilian space agency. President Eisenhower, a strong proponent of having civilians fly reconnaissance aircraft missions and, more generally, of portraying space as a peaceful realm, created NASA in 1958 to divert attention from the nation's other growing space program dominated by military and intelligence data-gathering concerns. In the civilian space effort, satellites would be used to gather weather data and to improve communications, and "manned" missions would explore outside Earth's atmosphere. These parallel programs, together with aircraft missions sponsored by both the military and NASA, have continued into the present.

Historians have identified a shift in the tenor of the military reconnaissance program following the creation of NASA and the departure of the Eisenhower administration. Under Eisenhower, the existence of a reconnaissance program was a fairly public activity—evidence of the nation's technological and military might. Under the subsequent Kennedy administration, this program became cloaked in secrecy. While the civilian space agency's achievements were much touted, its military and intelligence agency counterparts were not. CORONA, the first U.S. spy satellite, was secretly set into orbit in during this period. An outgrowth of the U.S. Air Force's Project Weapon System II7L, CORONA, like SAGE before it, instantiated the institutional relationships of the military-industrial-academic complex. It was built by Lockheed, containing components from the Fairchild Camera Corporation, and developed with the participation of a community of experts that included James Killian (formerly President Eisenhower's science adviser and then president of MIT), George Goddard (at that time the colonel directing the Photoreconaissance Lab at Wright Field, later Wright-Patterson Air Force Base), Bernard Schriever (at that time commander of the U.S. Air Force Western Development Division in Los Angeles, where he headed the ICBM program), Amrom Katz (chief physicist under Goddard's direction and later employed at RAND), as well other staff from the RAND Corporation.[13]

A planned focus of the classified reconnaissance satellite program was Earth surveys, both to enable more accurate targeting of intercontinental ballistic missiles (ICBMs) and to continually reassess the capabilities of the adversary. By comparing information gathered over time, analysts could plot increases in enemy arsenals or track the movement of weaponry around the globe. Surveillance of urban areas served targeting purposes and also to view the extent to which foreign nations were developing shelter programs for civilians and defense industries. As with so many other defense and aerospace innovations from the cold war, the development of CORONA and follow-on satellites such as ANNA-1B (launched in 1962) spawned much information that could be applied beyond the tasks for which it originally was created. For example, since engineers needed to verify the reliability of each new development in sensor technology aboard the satellite before it was used in foreign operations, they tested these innovations over territories in the United States, where topographical measurements and exact distances between landmarks already were well known (figs. 4 and 5). Since the Defense Mapping Agency did not operate a domestic mapping program, and given the close relationships earlier cultivated between military and intelligence agencies and the U.S. Geological Survey (USGS), CORONA's developers shared some of their data with the civilian agency. This had a significant consequence: the information was used to create more sophisticated maps of United States territory. Despite the secrecy of the reconnaissance program, the realization of its value for domestic mapping led to immediate technology transfer to the USGS, and in the late-1960s the agency moved its National Mapping Division to Reston, where it could have access to CORONA images.[14]

This partnership between military mapping and U.S. survey needs would be further formalized when the USGS and the Defense Mapping Agency decided to operate "joint committees and working groups to assure that map and chart specifications met all users' needs to the maximum extent possible."[15] Indeed, in the National Archives, a search for Defense Intelligence Agency images of U.S. territory will point researchers to a USGS collection. For many decades, then, the American history of aerial photographic surveys, both in academic and civilian government contexts, has been intimately intertwined with military and intelligence agency needs. In the words of one ge-

Figs. 4 and 5. Declassified images taken from the CORONA satellite show (fig. 4) the U.S. Pentagon in 1967 and (fig. 5) the Kremlin in Moscow, 1970. Each reveals the high-resolution imaging available to users with appropriate security clearance. Similar levels of detail about the urban landscape were unavailable to civilian users of the Landsat satellite. CORONA images courtesy USGS/EROS Data Center.

ographer there is a "correlation between the ultimate stupidity of man on the one hand and cartographic innovation on the other."[16]

NASA staff, too—like their colleagues at the USGS—worked in a civilian agency with continuing connections to the military space program. Although NASA had been chartered as a civilian agency by Eisenhower, NASA's cast of characters—from its staff to its contractors—made it impossible to miss the close and continuing ties to military and intelligence interests. Historian Howard McCurdy sees the origins of the space agency in several military and defense-related aeronautics organizations—the National Advisory Committee for Aeronautics (NACA), the Army Ballistic Missile Agency, the Naval Research Laboratory's International Geophysical Year, and the Air Force Ballistic Missile Program. Several installations such as the Marshall Space Flight Center, the Kennedy Space Center, and the Jet Propulsion Laboratory got their start as military facilities. The Marshall Space Flight Center (Huntsville, Alabama), which opened in 1960, was established as the Army Ordnance Guided Missile Center at Redstone Arsenal in 1950. The Kennedy Space Center (Brevard County, Florida), which opened in 1962, began as a missile launch range, the Joint Long Range Proving Grounds, in 1949. The Jet Propulsion Laboratory at Caltech (Pasadena, California), set up in 1944 for contract research for the Army Ballistic Missile Agency, became part of the space agency in 1958. Program offices such as the Space Nuclear Propulsion Office, a joint operation of NASA and the Atomic Energy Commission, were organizational declarations about the close ties between civilian and military interests in the space program. Major manufacturers of military aircraft and satellites—North American Aviation, Boeing, Douglas Aircraft, McDonnell Aircraft, Lockheed, Hughes, and TRW—expanded their contracts through ties to the civilian space agency. With the DoD and the CIA having served already for two decades as a customers for space technology before NASA's entry, national security priorities clearly had shaped the industry.[17]

Ties between the civilian space program and cold war space research and development for military needs included even more than facilities, contractors, and technologies. Historian Pamela Mack has described several prominent early figures at NASA as "on loan from the Army," formerly trained in military geography, or formerly working on defense contracts. For example, Keith Glennan, NASA's first administrator, successfully lured Wernher von Braun, the German rocket scientist, and his V-2 rocket team away from missile work for the U.S. Army to join the new space agency. U.S. Air Force General

Samuel Phillips was appointed to head the Apollo Program. TRW executive George Mueller was chosen to lead NASA's Office of Manned Space Flight from 1963, taking with him (as would the RAND men into the federal government) many air force officers with weapons system experience. And the Goddard Space Flight Center, established in 1959, was staffed largely by engineers and scientists from the Naval Research Laboratory, many of whom had experience working on the Vanguard Project.[18]

Thus, while the U.S. civilian space program under NASA was rhetorically distinguished from national security needs, in fact the two cannot entirely be disentangled. This was especially true for reconnaissance from air and space, where NASA missions occasionally served as fronts for classified intelligence-gathering. In March 1967, Lyndon Johnson—who while vice president to Kennedy had chaired the National Aeronautics and Space Council—would confess to a Nashville group of local government officials (off the record) that investments in civilian space exploration had yielded much military-relevant information. In his words, if nothing had come out of the space program beyond the knowledge that spacecraft surveillance provided, it still "would be worth ten times what the whole program has cost. Because tonight we know how many missiles the enemy has and, it turned out, our guesses were way off."[19] For these reasons, former McNamara defense aide Adam Yarmolinksy's 1971 study of *The Military Establishment: Its Effects on American Society* characterized NASA as one of America's three national security agencies, alongside the Department of Defense and the Atomic Energy Commission.

Space Age Management and Satellite Reconnaissance

The creation of an American aerospace industry followed closely on the heels of World War II. Rocketry for both air and ground-based weapons systems was central in wartime efforts and during the war offered opportunities for military contractors to innovate beyond aircraft production. Aircraft and missile companies such as Northrop, Lockheed, Bendix, and others found themselves important partners in military successes, and in the cold war period that followed each sought to maintain its competitive position.

The term *aerospace,* which dates to 1958, was a politically savvy choice for institutional self-promotion. Like the U.S. Air Force, which used the term to negotiate its position as the lead military agency in the early days of space research and development, the aircraft industry used the term to suggest how its experience in building aircraft for military reconnaissance as well as launch vehicles for missiles such as the ICBM well suited it to building military space-craft. (Modified rockets were used to launch satellites into orbit.) Companies such as Douglas Aircraft and Lockheed created new divisions for missiles and space research. Good relationships with military and federal agencies and a burgeoning space program with military and later civilian components helped the aerospace industry to grow quickly.

Yet successes in a climate of expanding spending on research and develop-ment for national security and a civilian space program became liabilities when political circumstances shifted. A downturn in aerospace spending ar-rived in the early-1960s, due in part to reductions in fear of the Soviet threat (President Kennedy signed a limited test-ban treaty in August 1963, and this was ratified by the U.S. Senate the following month) and in part to the pro-jected conclusion of the Apollo Program. Despite a national commitment to landing a man on the Moon by 1970, aerospace companies were concerned about their longevity. The pages of the trade weekly *Aviation Week and Space Technology* are filled with industry leaders' confessed anxieties.

Like their close cousins in defense research organizations, aerospace execu-tives and engineers decided that the survivability of their industry depended upon transferring their innovations to new markets. With a continuing push toward comprehensive planning in city governments, urban planning and management quickly emerged as targets of opportunity. Two specific industry products, "space age management" techniques and satellite reconnaissance technology, seemed especially suited for these tasks, each a tool to extend the reach of master planning as a means to deal with the complex problems of ur-ban life. The hope was that the cities deemed "ungovernable" might respond to a comprehensive planning effort mobilized on the order of the nation's space program.

The earliest aerospace technology-transfer efforts predated outbreaks of ur-ban violence. These disturbances, which seemed to result in part from a lack of scientifically sound planning, would provide further rationale for market expansion. Two NASA-sponsored researchers described how "difficulties of data acquisition may have delayed urban planning's attempts in facing some urban problems due to the imbalance between costs and benefits under the old methods."[20]

The aerospace community was far more muted than its defense industry colleagues when it came to using militaristic language to describe its potential interventions in city affairs. Yet the notion that the urban crisis was a national security crisis provided at least some of the motivation for their continued urban work. By 1967, Robert Hotz, the editor of *Aviation Week and Space Technology*, would draw the analogy between external and internal threats to speculate about the range of potential contributions to public service that the aerospace community might make. His article "The Turbulent Summer" suggested that the aerospace industry had a responsibility to respond to the nation's domestic challenges with the same vigor it had applied to protect Americans from foreign threats. Hotz proposed a variety of possible interventions, from transportation planning to job creation to riot control. Together, he suggested, these interventions "would absorb much of the energies now dissipated in violence."[21] If the urban crisis was a national security crisis, then the tools for intelligence gathering and analysis developed by the military and civilian space programs were ideally suited for application. The focus for transferring innovations became intelligence gathering and analysis, gaining a bird's eye view of city processes and executing a comprehensive plan.

Space Age Management

Space age management was not, like satellite reconnaissance technology, a deliberate contract project of NASA or the military space program; rather, it was the phrase that James Webb and others used to describe the amalgamation of managerial practices that had made possible the large-scale mobilization of people and resources required to create new space technologies. Webb, who served as head of NASA from 1961 to 1968, had been developing management expertise throughout his career in public service (he was a former head of the Municipal Manpower Commission, director of the U.S. Bureau of the Budget, and had been an undersecretary of state). During his time at NASA, Webb was credited with transforming the space agency into one of America's proudest administrative and technical achievements. He believed that both his "space age management approach" and the technologies it had helped to create could be disseminated widely—even more so than the knowledge gained from military research and development, where maintaining secrecy was of greater concern.[22] (While on many topics the aerospace and defense communities were of one mind, here Webb's comment about secrecy was likely a dig at his defense counterparts; Webb and McNamara repeatedly battled for territory and resources.)

The perception was that alongside the rapid development of successful weaponry and space-exploration technologies, NASA and its contractors simultaneously had developed a new approach to rational management, a set of techniques and an interdisciplinary approach to problem solving that might be applied to supervise any large-scale operation.[23] The "systems approach" was a central component, unsurprising given that the mother institution of systems analysis, the RAND Corporation, had its roots at Douglas Aircraft and that aerospace companies were early users of this analytic method. Webb acknowledged the failure of his predecessors, observing that "too often, the computer becomes the master of the systems disciple, rather than a useful tool in his hands" and that there were "consequently, serious difficulties in transforming into practice many of the rather nebulous concepts of the systems doctrine."[24] Yet given the speed of the space program's perceived success and the increasingly urgent problems faced by city governments, Webb and others were optimistic that their ideas could assist in realizing the much-touted comprehensive approach to planning change that had not yet been realized in U.S. cities and regions.

As early as the 1954 Housing Act, HUD had created a comprehensive planning grant program (section 701) to encourage cities to create master plans for urban change. The belief was that by organizing coordinated change on a large scale, the solutions to smaller and more specific problems would follow.[25] Yet as criticisms of urban renewal (including the Community Renewal Program) over a decade revealed, the comprehensiveness of planning practice versus policy—with the possible exception of large-scale demolition—was under question. Urban professionals' concern about how successfully to execute a comprehensive city plan paralleled a growing federal interest in comprehensive planning at the regional level.

Regional planning in a few metropolitan areas, including New York City and Los Angeles, dated to the 1920s. But it was not until the 1960s that the theme of regional interdependence became a fixture of federal-level reports on city problems. During the Nixon administration, numerous officials lent support to the regional planning ideal. For example, HUD Secretary George Romney spoke often of the integrated nature of urban and suburban problems in his many public statements about "the real city," proposing that city problems would never be solved without simultaneously dealing with suburban issues and calling for the dispersal of some low-income residents to the suburbs. The National Committee on Urban Growth Policy used aerial photography to showcase changing land use across America and to make clear that the legal boundaries of a city were not the endpoint of its urban problems. So, too, the 1971 report on urban problems issued by the U.S. Domestic Council described city problems as a regional matter. Observing that most land-use decisions were made locally but could have regional or national impact, the U.S. president proposed to provide money over a five-year period to encourage states and local governments to cooperate on regional land-use planning and environmental regulation.[26] Such findings would fuel the argument of *The Ungovernable City* (1977), the best-known statement on the integrated nature of urban and regional issues. Douglas Yates, the author, and the many city managers who used his "ungovernable city" phrase to describe their own experiences, observed that even effective policies in a city government might be undone by actions at the regional level and that intergovernmental cooperation was the only way to reduce conflicts among forces that often were working at cross purposes.[27]

Webb's enthusiastic promotion of space age management techniques in *Space Age Management: The Large-scale Approach* (1969) suggested how this more "systematic" and "scientific" approach might realize the goal of comprehensive master planning. His explanation was steeped in cybernetic imagery of communication, control, feedback, and homeostasis, and he drew an analogy between operating a space vehicle and running a city. Knowledge gained from the space program (for example, about how a satellite's internal motor systems reacted to gravity and centrifugal force) might offer insights into "the new kinds of management action and control systems we need to solve problems that we now face in urban life or to achieve economies of scale in programs for underdeveloped regions." The large-scale approach he imagined would have to "start with a sufficient committed input of power to achieve the equivalent of 'flying speed,' or 'orbital speed.'"[28]

Webb urged public managers to turn away from "government by crisis" to undertake a program of scientifically planned change. A staunch advocate of the military-industrial-academic complex, he believed that, given the opportunity, American universities and industries could help to coordinate the satisfaction of national needs. According to this view, the kind of large-scale mobilization that had created the Tennessee Valley Authority, fought World War II, and developed new techniques and technologies in the Manhattan Project, the ICBM program, and the space program might become a model for a coordinated national effort to command and control the direction of social change in U.S. cities and regions. In other words, by explaining how comprehensive planning already was part of the American style of management, he suggested how the techniques and technologies generated by the space age offered new means to achieve this kind of directed effort for urban change.

Even before the publication of Webb's 1969 book, in California, where a large portion of the state economy hinged on the aerospace and defense industries, a coalition mobilized to take action. Universities, aerospace companies, city managers, and the state's Democratic governor, Edmund "Pat" Brown, assumed a leading role to seek opportunities in the new economic climate for the California's many scientists, engineers, and aerospace institutions. A meeting in 1962 called by the Oakland city manager, Wayne Thompson, and a larger, follow-up conference the next year ("Space, Science, and Urban Life," held in Oakland and jointly sponsored by the Ford Foundation and NASA) got the conversation started. Participants included staff from Ford and NASA (including Webb). Also at the conference were researchers from think tanks and aerospace companies, including the Aspen Institute, the System Development Corporation, the Stanford Research Institute, General Dynamics, Douglas Aircraft, Lockheed, the Aerospace Corporation, JPL, and Northrop, and industry publication *Aviation Week* (later *Aviation Week and Space Technology*). There were also mayors, city managers, and representatives from urban organizations such as the League of California Cities and the U.S. Conference of Mayors and industry publication *American City*. Prominent intellectuals included Jerome Wiesner (MIT), Llewellyn M. K. Boelter (UCLA), Clark Kerr (president of the University of California system), and Martin Meyerson (director of the MIT-Harvard Joint Center for Urban Studies).[29]

Thompson, in 1963 still the Oakland city manager and, to boot, president of the International City Managers' Association, kicked off the meeting with a call to action that was often repeated throughout the decade—to "depart from time worn traditions and concepts and adopt space-age techniques to cope with the problems of our space-age cities."[30] Observing that many of the nation's intellectual resources, especially its scientists and engineers, had been focused on defense and aerospace challenges, he proposed a reorientation toward the problems of urban society. Discussions at this early conference about the most effective strategies for transferring aerospace innovations ranged widely, but the emphasis was on management, specifically systems analysis and the "space age management" of large-scale projects. Like the turn toward civil systems research at RAND, SDC, and MITRE, as the momentum built to put aerospace techniques and technologies to use in managing cities and cutting the costs of urban operations, there was a blossoming of opportunities to create and maintain social ties among disparate professional communities. At the 1963 meeting in Oakland and in similar fora that followed, sessions convened on topics such as "What immediate progress can be made to apply new space and scientific technology to greater use in our urban and industrial communities?" A mix of men came togather for discussion and debate, among them Richard Horner, a senior vice president at Northrop; Burnham Kelly, then dean of the School of Architecture at Cornell; William Pickering, director of the Jet Propulsion Laboratory; Samuel Silver, director of the Space Sciences Lab at the University of California, Berkeley; Robert Wood, then professor of economics at MIT; and Lewis Winnick, of the Ford Foundation (formerly chief of the Planning and Research Department of the New York City Housing and Redevelopment Commission and research director of New York City's Planning Commission).

In this climate, aerospace companies such as Lockheed, Hughes, and others began to seek opportunites for civil systems work. NASA's Sustaining University Program (SUP), established in 1961 to create university buy-in to the space effort and to help disseminate NASA spinoffs, undertook several projects focused on urban management.[31] At the University of Indiana, for example, an Aerospace Research Applications Center turned to transportation planning. At Drexel University in Philadelphia, in 1969 SUP funds, supplemented by contributions from the Kellogg Foundation, created an interdisciplinary Ph.D. program in urban management. And some wholly new organizations brought together older institutions to cultivate urban markets. For example, in 1966 General Bernard A. Schriever, former chief of the Air Force Systems Command and a developer of numerous missiles, created USA, Inc. (Urban Systems Associates, Inc.), an organization dedicated to attacking urban problems. Its affiliates included major players such as Lockheed, Northrop, Control Data, Emerson Electric, and Raytheon.[32]

Thus, years before NASA's Webb was invited to speak to the National League of Cities, a variety of constituencies already were busily finding ways to connect the immediate needs of NASA, the aerospace industry, and urban governments. Fears about dwindling contracts were transformed into opportunities by enterprising aerospace companies backed by state and federal government. Like RAND and SDC, institutions created to serve cold war needs, aerospace companies with extensive government contracting experience trumpeted how their managerial skills could meet the challenges of civil systems work in American cities.

Some aerospace professionals, including Tom Paine, NASA's deputy administrator and later agency head (who served on at least one HUD task force) drew upon the cybernetic worldview to argue that cities were information processing systems and that planning change required a cybernetic approach.[33] Others, among them James Webb, began from cybernetic themes and moved beyond them, shifting from an emphasis on cities as communication systems toward the language of cities as complex and large-scale systems. Aerospace executives such as Simon Ramo, vice chair of the board and chair of the policy committee at TRW (a former chief scientist for the air force's ICBM program), praised the systems approach and space age management techniques in an article in *Nation's Cities*. Reconnaissance experts such as Amrom Katz, who contributed to the design of one of CORONA's early cameras, declared that "NASA, as an organization that delivered the goods and performed so flawlessly on the MOM (Man-on-Moon) project, should realize that, in developing space hardware and activities, it also developed managerial skills and techniques that can be and perhaps should be turned to other big projects have little to do with space."[34] Management experts such as Dr. Leonard Sayles, of Columbia University's School of Business, observed "the fallacy of thinking that there is a clear distinction between essentially political-social programs on the one hand and engineering-technical-hard-science kinds of programs on the other."[35] Definitions of city systems as analogous to aerospace systems and the large-scale organization of cities as analogous to the large-scale organization of the aerospace industry passed over important distinctions. This wordplay helped to market the range of management innovations developed by NASA and its contractors; to use the title of Ramo's 1969 book, the aerospace community possessed the "cure for chaos."[36]

Satellite Reconnaissance

Satellite reconnaissance, another one of the space age's proud achievements, seemed an even more natural match for cities seeking to make comprehensive planning a reality. Just as enthusiasts for systems analysis and computer simulations promoted their innovations simultaneously as revolutionary and as simple extensions of existing practices, promoters at NASA and its contractors emphasized how their nonphotographic "remote sensors" were revolutionary improvements upon aerial photography, a tool transferred from military reconnaissance to city planning and management in the 1920s. (The term *remote sensing* was coined by Evelyn Pruitt of the Office of Naval Research in 1960.) Planning departments and other local government agencies were still using simple black-and-white aerial photography in a variety of capacities, and NASA officials speculated that these users would welcome the opportunity to employ advanced reconnaissance imagery of cities and of their surrounding regions, especially since satellite images were captured in digital form. Aware of city managers' interest in developing comprehensive urban information systems to complement redevelopment planning and the consequent push to digitize a range of survey data to introduce it into computers, they suggested that these images would soon take priority over the traditional, analog aerial photographic view. Continuous planning would be significantly easier when planners had the latest survey information at their fingertips.

Interest in applications of space age management came first, but as technology-transfer efforts and a civilian space program grew closer, interest in space-based surveillance for domestic needs began to gather force. Discussions about an Earth Resources Program date to 1964, when NASA administrators recognized what their colleagues developing CORONA already knew—that reconnaissance data of the domestic landscape similar to the kinds of intelligence gathered in other countries might be useful to the land management professions. NASA proposed creating an Earth Resources Technology Satellite (ERTS) to map Earth from space and gather data previously inaccessible to civilian users. Despite its continued use of CORONA imagery, the USGS (the primary mapmaker for local, state, and federal government), in the Department of the Interior, announced its own intention to create an Earth Resources Observation Satellite (EROS), and soon thereafter the two agencies established a partnership.

A few naysayers in the military and intelligence communities, such as RAND's Amrom Katz (later assistant director of the U.S. Arms Control and Disarmament Agency), opposed a civilian satellite program, recognizing that its images could yield military information. (He noted that while civilian and military space programs were artificially separated in the United States, there was no such separation in the Soviet Union.) Yet their concerns did not mean an end to the civilian satellite program. Beginning in 1972, a series of Earth Resources Technology Satellites (ERTS), as they were renamed, was launched. (This series would eventually become known as Landsat, and for the sake of simplicity and consistency, from hereon that name is used). With potential applications ranging from resource management to urban planning to geology to forestry, Landsat was poised to become the ultimate technology-transfer product for urban information gathering, created to showcase civilian applications of the U.S. space program and to generate continued public support for the space program in the years following the Moon landing.

As with the contemporaneous efforts to bring cybernetics and computer simulation to cities, during the six years of research and development leading to the first Landsat launch there was widespread speculation about its great potential as a scientific decision-making aid for local government planners and managers. Planning agencies, which generated relatively little of the data they used (they depended on collection and analysis from local, state, and federal agencies and the private sector) seemed likely future users of satellite data.[37] As promoters lauded the wealth of collected data to urban managers and planners, their terminology changed: *satellite reconnaissance* became *remote sensing; intelligence* became *inventories; surveil* became *survey.*

Planning for the early survey satellite in fact predated widespread urban rioting, but with the disturbances and the growing sense of a crisis of governance in American cities came additional urgency. At the Geography Branch of the U.S. Office of Naval Research, Robert Alexander described a growing consensus: declassification of military sensors would serve the national interest. Alexander urged that communication be "fostered between those who know the capabilities of the sensors and those who need desperately to obtain the data which the sensors can produce."[38] M. C. Branch—now professor of planning at USC—concurred: "It is no longer possible to follow the political dictum that 'nothing is accomplished until blood flows in the streets,' because that expression now has literal as well as figurative meaning." Municipal administrators would have to improve their response to the crisis in urban management to match the efficiency of other emergency services; otherwise, said Branch, "the dire prediction of some serious students of government that our larger cities are becoming unmanageable will indeed prove true."[39] Branch, a long-time supporter of continuous master planning, updated his 1948 book on aerial photography in city planning to address both the concerns of 1971 readers and the recent declassification of remote sensors. The techniques and technologies of the defense and space programs, he argued, could help to end the urban management crisis.

Five rationales were widely repeated in support of the value of satellite data in city planning: (1) satellite reconnaissance could offer a synoptic view, a total picture of the complex system in a given geographic area, ideal for comprehensive urban or regional planning; (2) satellites could provide repetitive coverage, a kind of longitudinal data series of change over time, offering continuous updates to the master plan; (3) satellite inventories were efficient methods of compiling information; faster and cheaper than ground surveys, they would be of special interest to managers of rapidly growing urban areas where most maps were out of date even before printing and distribution; (4) remote sensing was objective, presenting records of urban phenomena "unbiased by the planner's experience"; (5) satellites produced digital imagery that could easily be combined with emerging geographic information systems to match remotely sensed survey data with their ground-based equivalents. Remote sensing would not eliminate the need for ground surveys (ground surveys would to continue to provide supporting data and in fact were necessary for developing algorithms to make use of remote sensors); yet for cities already developing information systems in partnership with experts in systems analysis and computer simulations, the myriad benefits offered by satellite imagery would transform urban decision making.[40]

Complementing the space age management approach, reconnaissance technology would enable master planning at a new level. If cities and regions were constantly undergoing change, and if a systems perspective on cities identified this change as part of broader regional processes, only satellites could offer continuous large-scale monitoring, repeatedly updating digital imagery of land use. Promoters suggested inventories from space would have applications for decision making in comprehensive planning analysis, land-use inventories, socioeconomic and demographic studies, site and environmental planning, transportation and recreation planning, urban change detection, pollution detection, and resource management. Hype about the breadth of future benefits to be derived from the Earth Resources Satellite Program was so great—the applications to urban management were only one area of civilian benefit—that Jamie Whitten (D-Miss.) would ask colleagues at a House Appropriations Committee meeting, "Is it good for arthritis too?"[41]

In order to generate broad interest in its reconnaissance innovations, NASA created a series of programs to demonstrate how satellite imagery offered the solution to the problem of urban data for comprehensive planning that already was widely acknowledged. Beginning several years before the first Landsat was launched into space, NASA launched major efforts to reach city administrators with information about the benefits that its satellites could offer. One route that NASA chose, in partnership with agencies that included the National Science Foundation (NSF), the U.S. Census Bureau, and USGS, was indirect, seeding academic and government research to test applications of remote sensing in urban analysis. For example, at the University of Washington, a group was sponsored to explore the potential of remote sensing for land-use planning in the Puget Sound area, an administrative region where the Metropolitan Washington Council of Governments oversaw regional planning. This project, jointly funded by NASA and the USGS Geographic Applications Program, led to numerous published articles and at least one dissertation.[42] Researchers there and elsewhere used two strategies to test the validity of future satellite-based remotely sensed images. One was to interpret images of urban areas from spacecraft such as Gemini or Apollo—remotely sensed images not specifically gathered for planning purposes. Another was to run aircraft missions carrying nonphotographic sensors that would eventually be launched into space on Landsat.[43]

In 1971, NASA initiated its High Altitude Aircraft Program, using two Lockheed U-2C planes on loan from the air force. Based at the Ames Research Center, the Earth Resources Aircraft Program flew its first test flights on August 31 over San Francisco and Sacramento. To simulate the future Landsat orbit, five test sites were overflown on an eighteen-day repetitive schedule, with San Francisco/Sacramento and Los Angeles representing generic urban areas. The USGS Geographic Applications Program also sponsored high-altitude overflights to test multispectral scanners over twenty-six cities and four metropolitan regions.

In a closely related endeavor, the space agency hosted a variety of conferences to publicize findings from its sponsored research, helping to maintain the community of remote sensing specialists that the ONR had helped to create. The nation's best-established conference on remote sensing (held since 1962 and organized by the Environmental Research Institute of Michigan) was sponsored by the ONR. As ONR funding began to fall off in the later 1960s, NASA, in partnership with USGS, stepped up to the plate and became an official sponsor of the conference. Following Landsat's eventual launch from the Vandenberg Air Force base, the Goddard Space Flight Center hosted annual meetings to promote uses of the data this satellite collected.

A different approach to technology transfer engaged local government officials more directly. Several NASA-sponsored university teams surveyed state and local agencies about their existing sources and methods of data collection and how remote sensing might fit their information needs. A series of pilot programs tested applications of Landsat or simulated Landsat data to address local and regional planning concerns. One of the earliest was ARSIG (Applications of Remote Sensing to State and Local Governments in Arizona), which in summer 1972 brought together NASA, county planners, and the University of Arizona. Although the project's goal was to improve natural resource management in a rapidly urbanizing area, its actual findings emphasized natural resource planning and county-wide development planning, rather than urban issues. In the hope that such projects would expand the use of Landsat data in the city agencies that were developing geographic information systems, NASA's several regional applications centers served as clearinghouses of information.[44]

Thus, as their colleagues at defense research institutions promoted systems analysis and computer simulations as urban decision-making tools, the aerospace community emphasized space age management and satellite reconnaissance. Studies and applications of space age management and remote sensing in urban areas cannot be understood as emerging from an impartial set of scientific inquiries; rather, they should be read in the context of NASA history: for purposes of institutional maintenance, NASA had a growing interest in funding geographical research and applications built around aerospace innovations. In an era when cities increasingly were perceived as ungovernable, in an era when the Nixon administration was not a champion of NASA, a deliberate effort to promote aerospace techniques and technologies as a means to achieve "continuous master planning" for cities and even entire regions was a brilliant proposition.[45]

Mobilizing for Space Age Cities

Rhetoric from aerospace leaders and NASA circles focusing on the transferability of space age management and reconnaissance technology might have seemed like blue-sky fantasy had it not been echoed, for more than a decade, from several quarters in the federal government. Much as Lyndon Johnson had promoted the transfer of PPBS across federal agencies, his vice president, Hubert Humphrey, tirelessly praised aerospace innovations and their applicability to urban management in order to bolster support for projects that might extend beyond a single year's experiment. Humphrey's rhetoric, which echoed the language that Wayne Thompson had used several years earlier, went a step further, suggesting that the nation's defense and space agencies had a social conscience: in Humphrey's words, "maybe we're pioneering in space only to save ourselves on Earth."[46]

Federal backing for the transfer of management techniques and technologies, and indeed aerospace experts themselves, to urban problem solving in state and local government came as early as 1965. That year, Senator Gaylord Nelson introduced a scientific manpower utilization bill (S. 2662), proposing that $125 million in federal funds be provided for state and local governments to use systems analysis "to design computer programs that would test various solutions for urban woes with the same logic used in designing moon rockets."[47] Federal agencies including HUD and the Department of Labor (DoL), independently or in collaboration with NASA, followed NASA's lead, mobilizing resources to implement their own targeted efforts to profit from the space age.

In 1966, HUD and the White House Office of Science and Technology co-sponsored a three-week summer study, "Science and Urban Development," at the National Academy of Sciences Study Center at Woods Hole, Massachusetts. Summer studies were a regular feature of research at DoD and NASA; this meeting was the first of its kind for officials from HUD. With Robert Wood, of HUD, and Walter Rosenblith, of MIT, as co-chairs, scientists and engineers from universities and the defense and aerospace community were brought together with urban specialists to discuss new approaches to city problems. The study group's findings were published in the report *Science and the City* (1966), crafted by Volta Torrey, a NASA researcher loaned to HUD for the project. Throughout Torrey's report, spaceships and cities, the Moon program and the Model Cities program, are juxtaposed to reveal their striking differences and to mobilize support for the proposition that urban programs follow the space program's lead. Torrey communicated the study group's proposal that HUD model its future urban development efforts after NASA and "orchestrate" the efforts of intellectuals with the public and private sectors so that all parties could find benefits in a joint effort to address urban problems.[48]

Despite the study group's expression of pessimism about the past, they saw hope in some of the Johnson administration's proposals for urban experimentation. For example, the Model Cities Program appeared to be designed as a scientific experiment in master planning, with individual local efforts serving a function "analogous to those of the astronauts' pre-Apollo flights."[49] Careful study of such pilot projects, like the careful analyses of NASA test flights, would help to direct HUD's efforts toward more successful future outcomes. Four decades earlier, the Chicago School of urban sociologists had deemed their city a laboratory as they sought to formulate the scientific laws behind urban change; now, for the defense and aerospace communities, cities across the nation had become laboratories for social experimentation.

HUD's interest in comprehensive planning and space age housing and urban development continued full throttle into the Nixon administration, defying political partisanship. In 1969, the transfer of aerospace manpower expertise gained recognition at a high level of government when HUD Secretary Romney appointed Harold Finger, formerly of NASA and a registered Democrat, to be the first assistant secretary of HUD in charge of urban research and technology.[50] Finger, who had worked at NASA from the agency's first day of operation, had risen up the chain of command to become manager of the Space Nuclear Propulsion Office, operated jointly with the Atomic Energy Commission, then associate administrator of NASA, the agency's fourth highest office.

In his three years at HUD, Finger directed the agency project that best embodied efforts to bring housing and urban development into the space age: Operation Breakthrough. Operation Breakthrough aimed to promote large-scale production of prefabricated housing and model communities to encourage more efficient and economical solutions to the nation's housing needs. Given that the rapid manufacture of defense housing during World War II had been a success, Operation Breakthrough proposed prefab housing to remedy some of the acknowledged failures of urban renewal—among them, continuing housing shortages and an overconcentration of poor and minority residents in inner-city high-rises. To use Webb's terms, the goal was to reach "flying speed," or "orbital speed," through contracts with industry, and chart a new course for success in housing planning.

Webb's 1972 description of the rationale for and projected successes of Operation Breakthrough is a classic example of the rhetoric of his day, and how the ideas he theorized about in his 1969 book seemed to be finding application in federal urban programs. He explained his take on the situation when asked about Operation Breakthrough at a conference: Finger, he said, had applied to housing development "the most successful techniques he learned in building rockets" and, more generally, in working with NASA's management systems. "It's going to produce some major forward thrust in the field of housing," Webb concluded—"thrust in terms of how you get it done, in the operational requirements and the nonoperational requirements."[51] Webb's analogy to the space program concealed one important distinction: that when Finger was at NASA, his R&D budget was $4.5 billion; at HUD, it was $11 million. Although this was HUD's largest-ever budget—larger, in fact, than all previous federal housing budgets combined—it was only a small fraction of the funds available in his earlier job.

NASA, USGS, and HUD were not the only federal agencies interested in transferring aerospace expertise to urban management with aid from university partners. During the 1970s, alongside manpower initiatives created to address the perceived causes of urban violence, another kind of manpower-training program came along. With support from DoL and HUD, universities were asked to work with civic organizations to retrain aerospace workers for jobs in urban management. DoL and HUD contracted with the National League of Cities and the U.S. Conference of Mayors (NLC/USCM) to administer two manpower programs: Project ADAPT (Aerospace and Defense Adaptation to Public Technology) and the Joint Aerospace Employment Project. Both aimed to orient aerospace workers leaving that industry to priorities and challenges in local government. The NLC/USCM partners asked two institutions with distinguished records in both aerospace and urban research to assist, and during the summer of 1971 scholars from MIT and the University of California at Berkeley conducted month-long crash courses in urban studies.[52]

At MIT, Project ADAPT enlisted faculty at the Department of Urban Studies and Planning (although the project title at MIT included the word *defense*, the emphasis, as at Berkeley, was on aerospace). The most prominent participants at MIT were Francis Ventre (project director), Larry Sullivan (assistant director), and Lloyd Rodwin, the department head. At Berkeley, the Aerospace Employment Project was run through the University's Extension Program, with assistance from the College of Environmental Design. Workshops were organized by William Wheaton, at that time dean of the College of Environmental Design. Wheaton, a long-time promoter of defense concerns in urban government, had served on the 1950s research team for Project East River while a professor of regional planning at Harvard; he had also worked at the federal housing agency and been a U.S. delegate to the United Nation's Economic and Social Council for housing and related issues. In the Berkeley program, mornings were dedicated to lectures; afternoons and evenings included field trips, urban gaming simulations, and special events that included, for example, "rap sessions" with mayors (the mayors of Cleveland and San Francisco participated).

It was only a few years since the National League of Cities meeting in 1966 had put James Webb on the defensive. Now the leadership of the league and the U.S. Conference of Mayors were singing a different tune: "If you can't beat 'em, join 'em." Program organizers imagined that aerospace professionals would be best suited to jobs in mayors' offices, central budgeting, city planning, or Model Cities agencies. MIT, in particular, tried to prime the market for the arrival of its recruits. Program organizers there ran a placement bureau, contracting with municipal leagues in Ohio, Texas, Georgia, Michigan, and Pennsylvania to develop jobs and job-matching programs.[53]

The National League of Cities and the U.S. Conference of Mayors were just two of the many organizations of urban professionals that perceived potential benefits in a partnership with aerospace. Early successes of NASA's main technology-transfer initiative, the Technology Utilization Program (TUP), led to discussions about creating a joint "NASA/City program." As a result, in late-1970 a partnership between NASA and the International City Management Association (ICMA, a proponent of systems analysis and computerization in urban government) became a reality. Its focus was on the application of technology to urban problem solving. The organization's first meeting, that same year, gathered aerospace executives and representatives of eighty cities at the Kennedy Space Center. In January 1972, the program changed its name to become Public Technology, Inc. (PTI), expanding its mission to include technology in state and local government and expanding its membership to include ICMA, the Council of State Governments, the National Association of Counties, the National Governors' Conference, the National League of Cities, and the U.S. Conference of Mayors.

PTI joined with the American Institute of Aeronautics and Astronautics (AIAA, which later partnered with NASA) to host a series of conferences on technology transfer to urban areas, bringing together city managers and aerospace executives with representatives of NSF and NASA throughout the 1970s. The PTI/AIAA partnership also collaborated to administer an NSF-sponsored program called the Urban Technology System (UTS), focused on improving the delivery of local services while cutting costs. "Technology agents" in specific areas were charged with implementing a variety of management innovations in energy, public works, public safety, and other areas; these included cable television, automated information systems for policing, geographic information systems, and computerized analysis of fire station location. Publications such as *UTS Briefs* made explicit the recognition that local conditions sometimes impeded the transfer of innovation from one locale to another; nevertheless, they expressed the conviction that it was worthwhile to try.[54]

Once again, experts from outside the urban professions were leading the charge to try a new approach to tackling urban problems. Chapter 5 considers the results of their mobilization efforts for two American cities. Again, close examination of the experience of New York City and Los Angeles finds an outcome that better satisfied the interests of the technology-transfer enthusiasts than it addressed the needs of urban residents.

CHAPTER FIVE

Moon-Shot Management for American Cities

In 1968, William Mitchel, deputy assistant secretary for management systems at the U.S. Department of Health, Education, and Welfare, attended a workshop on technology transfer for local government leaders organized by McDonnell Douglas. Such meetings were commonplace in this era—organized efforts to bridge the acknowledged cultural chasm that separated the aerospace industry from the local governments it aimed to assist. Despite their best efforts to find common ground, however, Mitchel surmised the hosts and their guests were speaking past each other. He concluded that America's aerospace community and its urban professionals—like the nation's black and white citizenry—remained "two worlds separated by values, infrastructure, language, skills, and economic resources" into "haves" and "have nots."[1] In reflections that echoed the Kerner Commission's description of a nation divided, Mitchel predicted a bumpy road ahead for future collaborations.

Mitchel's assessment of the workshop was a minority view, however, in a climate of enthusiasm for space age approaches to city administration. Several years would pass before such skeptical assessments would become mainstream. The 1969 Moon landing extended support for the variety of programs that NASA had created to showcase how innovations developed to address specific challenges in space also had relevance to problems on Earth. Through the 1970s, a variety of individual and institutional actors continued to rally around the idea that space age innovations had the potential to save U.S. cities. Alongside the more recent importation of other military analytic tools, nearly fifty years of aerial photography and airphoto interpretation across city agencies—themselves products of military research and development—made many observers especially optimistic that two specific products of the space age—space age management and satellite reconnaissance—would be natural complements to comprehensive urban planning. Local government officials actively recruited aerospace executives and engineers to work for their cities, creating new programs and bureaus to experiment with these "cures for chaos."

From the vantage point of 1968, Hubert Humphrey and William Mitchel placed competing bets on the future outcomes of collaborations between aerospace and urban government. Where Humphrey predicted a new generation of space age cities, Mitchel forecast little change ahead. Neither was to find himself completely correct. The aerospace community did significantly influence the history of city operations, but not in the precise ways it had planned. Despite formal steps by NASA and its industrial allies to bring space age management and satellite reconnaissance into widespread civilian use, city administrators instead chose to adopt a different set of aerospace techniques and technologies, little promoted, to address their intelligence gathering and analysis needs. Advances in techniques for photointerpretation and technologies for aerial surveying became staples of comprehensive planning and neighborhood revitalization projects, used alone or integrated into geographic information systems. To comprehend why city planners and managers of the 1960s and 1970s preferred these innovations requires a nuanced understanding of how each technique and technology fit—or did not fit—the larger history of "values, infrastructure, language, skills, and economic resources" that defined intelligence gathering and analysis in U.S. cities.

From Strike Planning to City Planning

This book began with an account of Americans' fears about aerial attack and how in the years following World War II those fears became rationales for dispersal planning. In fact, as early as the 1920s city administrators already were well acquainted with the techniques and technologies that would make possible the accurate targeting of Hiroshima and Nagasaki, applying innovations in aerial reconnaissance to their work. The growth of interest in aerial photography for civilian purposes during the 1920s and 1930s would not have been possible without the expertise gained from earlier military conflict. Many of the men originally trained for wartime service in World War I became agents of technology transfer, finding civilian and further military applications for their talents in the years that followed. For example, George Goddard, who served as an instructor in photographic interpretation at the U.S. Army's photography school at Cornell University and organized the first Army Aerial Photographic Mapping Unit, went on to work for the Federal Board of Surveys and Maps. By 1946, Goddard was back in military and intelligence operations, deputized with recording the Operation Crossroads atomic tests on film.[2] Similarly, Sherman Fairchild, whose Fairchild Aerial Camera Corporation manufactured the army's standard camera for aerial photography in the 1920s, expanded his market for surveying cameras through contracts with the U.S. Coast and Geodetic Survey in the 1930s. Fairchild went on to create several more companies and to supply the U.S. armed forces with numerous standard aircraft and aircraft-based survey technologies, including a photo-reconnaissance system for the B-58 bomber.

During the interwar period, several private aerial survey companies—Fairchild Aerial Surveys, Ansco, General Anline and Film Corp., Abrams Aerial Survey Corporation, Aero Service Corporation, and Carl Zeiss, Inc.—were established to profit from growing interest in aerial photography for mapmaking, land-use studies, city planning, and engineering projects. With the creation of specialized companies, surveys and mapping techniques originally developed for military users became more easily transferable to civilian customers in government and the private sector. From the 1920s, academic researchers in geography and in civil engineering began to work with aerial surveys.[3] The industry further professionalized with the founding in 1934 of the American Society of Photogrammetry, whose conferences brought together military officers, aircraft industry representatives, academics, and government officials.

Almost as soon as the first aerial survey companies were created, states and municipalities began to enlist their services to create photographic records for administrative use. The city of Middletown, Connecticut, for example, contracted with New York–based Fairchild Aerial Surveys in the 1920s to conduct an aerial survey for tax purposes. Beginning in the 1920s, Aero Service Corporation shot numerous pictures for the Philadelphia Tri-State District, and in the early-1930s mapped all of New Jersey. Indeed, the popularity of aerial photography as a survey method maps nicely onto the history of urbanization in the United States: surveyors used the changing urban landscape to urge municipalities to take stock of their constantly changing assets.[4]

Like advertisements for many new technologies, advertisements for aerial surveys served both commercial and educational purposes. The ads counseled users about the benefits of different scales for different needs, pointing out the advantages of aerial images over traditional ground surveys. Private companies created ads, pages long, to explain the value of the new industry for city planning and management and to show city managers what their colleagues in other municipalities were up to. Benefits of the new technology were said to include detail at the level of 100 percent, perfect scale, low cost, efficient production of maps (in weeks, not years), and maps that could be more easily read by average citizens. The point about legibility for citizens was frequently argued because of its value in settling conflicts—despite the fact that "photo-interpretation" was an area of growing professionalization. For example, in disputes over tax equalization, few citizens could quibble with the validity of aerial images. Experience from several places confirmed that "the mental protest of most citizens upon payment of their taxes" would be eliminated when the fairness of taxation surveys could easily be viewed.[5]

By the end of the 1920s, among a range of "typical uses for Fairchild Aerial Surveys" cited in an advertisement were tax maps, municipal maps for planning, zoning, regional planning, property record maps for tax assessment and property appraisal, and utilities planning. Many additional uses were listed in an April 1928 Fairchild ad in *Connecticut Industry* that carried the headline "Solving Other Municipal Problems with Fairchild Aerial Surveys": engineering studies of power distribution, right-of-way maps for highways, location surveys for transmission lines and railroads, property surveys, traffic control, developing town plans, locating reservoir sites, park layout, construction record maps, and more. Diverse applications were also listed outside the industry's promotional literature; for example, a 1936 report from the Mississippi State Planning Commission testified its aerial survey program was used in tax assessment and seventeen other tasks.[6] Thus, urban planners and managers have long included aerial photography in their arsenal of approaches to information gathering.

Interest in aerial photography at the municipal level was bolstered by federal support during the New Deal. The U.S. Department of Agriculture (USDA) called for aerial surveys to map farmland nationwide, creating a photographic encyclopedia that was available to states and local governments. Regional planning organizations such as the Tennessee Valley Authority employed the technology in resource studies. During the 1940s, the USGS began to release a series of maps of the United States overlaid with county-by-county information about available photographic records from private surveyors and government agencies. Also beginning in the 1940s, the U.S. Census Bureau, an innovative user of new technology throughout its long history, explored the use of aerial photography as a way to reduce census undercounts, particularly in rural areas.[7]

Building on military photointerpretation techniques and reconnaissance technologies, civilians sought new applications for aerial imagery. Pioneering photointerpretation work from Norman Green and Robert Monier in the air force's POP KEY program, for example, had linked the characteristics of physical environments to socioeconomic information. Beginning in the 1950s, academic geographers, sociologists, and even some city planners began to apply these techniques in urban renewal planning. Housing studies were a focus for much of their work. Matthew Witenstein, Leo Silberman, Roger McCoy, and Ernest Metvier, as well as Green and Monier in their later civilian studies, were among scholars who pioneered indices to relate housing information (e.g., lot size and density, and the condition of residential streets) with socioeconomic status. These early studies were unable to develop a single algorithm that could easily be applied in multiple studies, but in contrast with the time-consuming and costly efforts of ground surveys and interview questionnaires, the aerial information, appropriately interpreted, yielded similar results in less time and for less money. It also offered a record of changes over time.[8]

Alongside these academic efforts, similar approaches were making inroads into local planning. For example, in the 1950s the Housing Association of the Delaware Valley (formerly the Philadelphia Housing Association) made use of Aero Service Corporation aerial photographs to assess the conditions of urban blight in their area; and beginning in the 1960s, the Boston Redevelopment Authority contracted with two companies, Fairchild Aerial Surveys and Lockwood, Kessler and Bartlett, to aerially map the city for redevelopment planning. In Houston, public-health surveys conducted during the 1970s discovered that remotely sensed environmental surveys were as effective as census data for generating information about public health in the city.[9]

The movement of reconnaissance technologies from military to urban use followed the export of interpretive techniques. Until the late-1950s, the two major technological innovations in military reconnaissance from this period, nonphotographic sensors and spacecraft surveillance, remained classified and inaccessible to civilian users. As a result, in the decade following World War II, states and cities persisted in using standard black-and-white aerial photography for surveys and mapmaking. A New Hampshire State Planning and Development Commission report from 1949 summarized existing photographs of the state at close range (the most distant being 1 inch to 5280 feet) and proposed creating a small state agency (two people) to procure and distribute such photos. New Hampshire legislators drafted this idea into official state policy, encouraging their state to create an aerial survey and an aerial photography division under the state's Planning and Development Commission. Similarly, in Pennsylvania, a state law passed in 1951 made aerial photography part of the legal process for assessing taxes in that state.[10]

State and local governments continued to praise black-and-white aerial photography for its efficiency and low cost. In 1949, Cleveland officials claimed that aerial mapping was saving taxpayers $2 million. Abrams Aerial Survey Corporation ads from 1952 noted that, compared with a ground survey, aerial surveys required just one-fifth of the time and expense. Savings were especially large when local governments made use of federal or state aerial surveys. In such cases, local agencies did not have to contract for overflights or have a resident staff of specialist photointerpreters; rather, as one publication from USDA noted, with minimal training and access to government airphotos from federal, state, and local agencies or private companies, local governments could use existing aerial surveys for negligible cost. But even if a new aerial survey had to be made, the conclusion was still the same: "The cost of having new airphotos made for this purpose is more than offset by the overall savings."[11]

With the declassification of nonphotographic sensing methods such as color infrared (CIR) photography, thermal imaging, radar, and multispectral sensing in the late-1950s and early-1960s came a potential technological turning point for civilian aerial surveying. Much as in the postwar period following World War I, a community of experts became part of the technology transfer, moving from military to civilian settings. Kirk Stone, for example, who had served as a geographer in the Office of Strategic Service, would pursue research on aerial photography in civilian urban analysis. Similarly, Robert N. Colwell, a University of California–Berkeley forestry professor, who had worked with camouflage detection film in his classified research for military sponsors (part of air force research for the SAMOS satellite), went on to work with color infrared film, its civilian analog. Colwell eventually headed Berkeley's Space Sciences Lab and then NASA's forestry group, promoting the uses of Landsat in natural resource planning. Another pattern was repeated from the World War I experience: In 1922, Willis Lee's *The Face of the Earth as Seen from the Air* had expressed the debt of civilian applications to the recent war's innovations in military technology; in the 1960s, many primers on civilian photointerpretation—for example, T. Eugene Avery's *Interpretation of Aerial Photographs* (1968), gave their due to World War II. In Avery's book, the final two pages are devoted to aerial photos of Hiroshima, showing the city before and after the dropping of the bomb.

By the mid-1960s, then, alongside discussions initiating a satellite program for Earth surveys, the field of photointerpretation enjoyed a period of expansion. A turn toward interpretations that stressed social, as well as physical, details renewed interest in planning applications for the decades-old aerial survey method, an interest that was reinforced by the availability of newly declassified sensor technologies. When NASA offered funding for geographical research and applied pilot programs in the late-1960s (part of efforts to stimulate interest in Landsat), both academic geographers and city administrators were widely familiar with aerial surveys. But there was a sharp contrast between how these two communities embraced the new opportunity for technology transfer. While academic geographers were captivated by remote sensing, few city administrators expressed interest in changing their established practices. The remote sensors that had recently been declassified were not of a quality that would make government satellites preferable to private aircraft as platforms for city surveys. The histories of space age management and satellite reconnaissance in New York City and Los Angeles illustrate how city officials, initially interested in the products of the space age, eventually discarded most of them.

New York City

Any history of linkages between aeronautics and urban government in New York City has to highlight Sherman Fairchild and John Lindsay.

Fairchild, a major figure in the history of aerial reconnaissance (both military and civilian), brought aerial photography into local use in the city from the 1920s, when he established Fairchild Aerial Surveys to seek contracts with civilian clients. Four decades later, during his seven years as mayor, Lindsay attempted to take the city into the space age. At the same time that he was luring the RAND Corporation into a partnership with the city, Lindsay also was cultivating ties with aerospace experts from industry and from NASA. These two, separate, efforts at managerial innovation based on the aeronautics experience met with very different long-term outcomes, and the New York City experience matched the fate of similar efforts across the nation: that is, aerial photography maintained its popularity for decades, whereas space age management proved to be a short-lived fad.

Sherman Fairchild's Aerial Survey Company first mapped the entire city of New York in 1922. Within weeks of his overflights, numerous customers were making use of the photos, among them the NYC Board of Estimate and Apportionment (the city planning department of that era), the NYC Financial Department, the City of Newark, the Consolidated Gas Company, the New York Telephone Company, the Corn Exchange Bank, the Erie Railroad, the Transit Commission of the State of New York, the Public Service Electric Corporation, the New York Edison Company, the Borough of Queens, the Erie Railroad, and several public utilities. A tireless promoter, Fairchild described his work in such diverse publications as *Scientific American* and *Connecticut Industry;* in each, he detailed how, as a result of aerial surveys, government agencies and private companies across the country already better understood their changing cities' landscape.[12]

Early testimonials praised the new survey method for the amount of time and money saved—even accounting for time lost waiting for clear weather and the preparatory work that had to be done on the ground. Nelson Lewis, a former chief engineer of the NYC Board of Estimate and Apportionment who in 1923 was director of the Physical Survey for the Regional Plan of New York and Its Environs, pointed out that it now took just days, instead of months, to take stock of the city. In other cities and towns, given the revisions to the tax rolls that aerial surveys produced, the method paid for itself within a few years. Interest in the new perspectives on the city afforded by aerial views became so widespread that several of Fairchild's images of New York were reprinted in the magazine section of the *New York Times.* In a long feature article in 1928 describing the city's changing landscape, author H. I. Brock speculated that builders would soon build so that structures would appeal to viewers from the air. From the days of the earliest surveys, the low-altitude flights and high-quality image resolution made for sharp views. In 1922, at the time of his first city survey, Fairchild described how "Mrs. Smith's washing can be seen hanging on the line, so you know Mrs. Smith's wash day even if you don't know Mrs. Smith."[13]

Fairchild's company did not hold a monopoly in surveying New York City from the air. Other surveyors included the Perkins-Elmer Corporation (a major CIA contractor), Skyviews, the McLaughlin Air Service, the Air Map Corporation of America, and the Aerospace Corporation. They took photographs on contract for municipal government agencies such as the NYC Department of Parks. They also surveyed the landscape for real-estate developers and brokers and created photographic records for retail and industrial clients who were deciding on where to locate businesses (figs. 6 and 7).[14]

Documenting the numerous ways in which aerial surveys were being used in urban areas was difficult, Ashraf Manji noted in a 1968 study for NASA; while widely used, aerial photography was rarely the subject of formal reports. Thus, past applications of this method in most cases must be uncovered indirectly. In New York City, the extent of the use of aerial photos is best found by examining planning studies. For example, in 1922 the *Municipal Engineers Journal* ran an article on traffic congestion, featuring aerial views of the city, which suggests that the technology had made inroads into transportation planning; however, another traffic survey, from 1929, reported in the *New York Sun,* used only ground surveys. And while a 1939 study of the East Midtown District for the East Midtown Planning Association featured a Fairchild aerial survey of the district on its cover, the city's master plan that was released in 1940 depended on drawn maps and ground surveys.

Aerial surveys became increasingly popular over time. Airphotos are scattered throughout documents related to Robert Moses's extensive stamp on the

Figs. 6 and 7. These airphotos of New York City, created decades before satellite imaging, show why city planners and managers preferred aerial surveys even when satellite imaging became available. Figure 6, prepared by Fairchild Aerial Surveys during the 1920s, shows much of Manhattan. Figure 7, prepared for the NYC Department of Public Works, offers a close-up of a single neighborhood, Battery Park, 1940 (the surveyor's identity is not known). Such aerial images were widely used by urban planners and private businesses for development purposes. Reprinted with permission of the Municipal Archives, Department of Records and Information Services, City of New York.

city, from a 1939 Department of Parks report on the future of Jamaica Bay to a study on civil defense planning, to many of the reports on slum clearance from a committee that Moses chaired, beginning in 1950. Still more aerial images can be found in documents from the city's Community Renewal Program, one of which described residential "reconnaissance" studies by surveyors from the planning department. These studies combined aerial photographs with maps and detailed ground surveys. On some occasions, New York City's planners, like the military planners described in the preceding chapter, even took to the air: the chair of the City Planning Commission, Donald Elliott, flew over over parts of Brooklyn and Queens, for example, in a police helicopter as part of a planning study.[15]

The standard that Sherman Fairchild set for aerial surveys in the 1920s was long-lasting. When the City Planning Commission prepared its master plan for 1969, the document's foundation was black-and-white aerial photography juxtaposed with drawn maps, social documentary photos, and statistical data. This six-volume document outlined the commission's plan of attack for directed change based on detailed intelligence about its target: the city's neighborhoods.

A Geographic Information System for New York City

City planning in New York City continued to use black-and-white aerial photos from the late-1970s and into the 1980s. The major technological shift to follow black-and-white aerial surveys was not a leap to nonphotographic sensors or spacecraft surveillance; rather, it was a move to integrate aerial photographic information into the city's developing computer systems. Until the late-1960s, information gleaned from aerial photographs was maintained separately from ground survey data. These efforts were the beginnings of civilian geographic information systems (GIS), which according to John Cloud and Keith Clarke have military origins in TOP SECRET defense and aerospace mapping projects, including the SAGE air defense system and the CORONA reconnaissance satellite, as well as other members of the military geographic intelligence system family (for example the Province Hamlet Plot in the Vietnam War).[16] Geographic information systems differed from other urban information systems in their ability to overlay aerial photographic information on map displays and to link data from ground surveys to specific locations.

Records in the New York City archives show a turning point in 1966, when alongside his efforts to court RAND's defense management experts Mayor

Lindsay began to cultivate relationships with aerospace executives and engineers. The goal was to bring space age management to the city. That year, Lindsay established a Science and Technology Advisory Council (also known as the Committee on Science and Technology), comprised of industry and academic experts in the New York metropolitan region, and created a Management Science Unit in the Mayor's Office of Administration, the first of its kind in the nation. In July 1967, Dr. Emanuel "Steve" Savas became head of the new unit, with the rank of deputy city administrator.

Savas came to the city with expertise in computer applications and operations research. As manager for Urban Systems at IBM following army service in the Korean War and as general chair for the ACM conference on the Applications of Computers to the Problems of Urban Society, Savas already had been thinking about how managerial innovations with military roots could aid the administration of city government. The new science of management would allow public administrators "a powerful but benign weapon with which to reassert control and direction" in the struggle to save American cities.[17] Savas's background, his description of management science in military terms, and his enthusiasm for systems analysis and computer modeling identified him closely with the community of defense intellectuals from CONSAD and RAND at work for the city. Yet Savas sought a different set of projects for the Management Science Unit, among them a GIS to assist in coordinating the sharing of information across city agencies. This system would be the first to integrate aerial survey data with ground-based information for physical and social planning in New York City.

Thus, when in March 1969 Lindsay announced the city's intention to computerize land and building records, a geographic information system known as GIST was under development in the Management Science Unit of the Office of Administration, the project headed by Timothy Costello (deputy mayor and city administrator) and Robert Amsterdam (GIS director for the Office of the Mayor). Other administrators (at the Department of City Planning, the Finance Administration, the Housing and Development Agency, and the Bureau of the Budget) also played a role. GIST was able to integrate information from multiple city databases to prepare analyses down to the level of a city block or parcel. In its earliest stages, baseline information was provided by maps from the Department of City Planning, the City Planning Address Coding Guide, and Automatic Location Table (AULT) Land Detail Files created from aerial photographs taken for the Tri-State Transportation Commission in

1962. All of these data had to be translated into machine-readable form so that a program called SYMAP could generate the relvant maps.[18]

SYMAP (Synagraphic Mapping System) was an automated mapping program developed to be a general-purpose system with transferability. The software was first developed by Howard Fisher, of the Harvard Graduate School of Design's Laboratory of Computer Graphics and Spatial Analysis (an early center of GIS research, established with support from the Ford Foundation; much of its research was sponsored by the Office of Naval Research and other military sources). SYMAP was the pioneering system of its kind. A mapping program that displayed information based on census and administrative divisions, this Harvard-created program was not a surprising choice for New York City, given that the deputy director of planning, Philip Wallick, had in May 1967 taken an eleven-day course on computer mapping at the Harvard laboratory where it originated.[19]

In contrast to many of the computer models being developed in city governments at this time, GIST actually was used in administrative decision making. By January 1970, the Department of Social Services adopted the system to analyze caseloads and to determine the location for a new satellite center. The Finance Administration had used it to analyze real-estate transactions in the city and to determine property taxes. The Office of Civil Defense had applied it to decision making for shelter planning. The Office of Administration had employed it in an election districting study. A year later, users included the city's Health and Hospitals Corporation and the Department of Traffic; the system was also used in planning lunch programs in school districts. GIST developers aimed to create a database that all city agencies eventually could tap into from remote locations. This was precisely the type of urban innovation that had aroused the interest of NASA administrators since, with only minor adjustments, satellite data could be integrated into the system. Yet the evidence available suggests that despite increasingly close ties between city administrators and the aerospace community, GIST developers expressed no interest in taking their system in that direction.[20]

NASA and New York City

As early as May 1968, NASA invited Lindsay to attend a conference of aerospace bigwigs, including NASA administrators and the commanders in chief of North American Rockwell, TRW, United Aircraft, McDonnell Douglas, Goodyear Aerospace, and Boeing, to discuss potential urban applications of aerospace innovations. Lindsay—the only mayor invited—sent Timothy Costello in his stead, and Costello reported back that while presentations from Aerojet, TRW, and Lockheed "were relatively elementary systems applications," he had had "an opportunity to establish useful relations both with the Industry and with NASA, represented by its Deputy Administrator Tom Paine." He also had had "a brief opportunity to present the city's needs to the Aerospace executives and following that, to NASA."[21] The cultivation of these relationships continued, and on December 9, 1971, several city agencies, including the Mayor's Office and Budget Bureau, met with NASA's assistant administrator for technology applications.[22]

To mobilize further support for technological applications, in 1972 Lindsay created an Office of Science and Technology in the administrative division of the Office of the Mayor. Its director, Leonard Naphtali, a deputy city administrator, earlier had been hired to the city from Mauchly Associates, a computer firm run by computer pioneer John Mauchly.[23] In a memo, Lindsay explained why he had created yet another office, separate from the Management Science Unit and the New York City RAND Institute, oriented toward cultivating relationships with the defense and aerospace communities. Its purpose, he told administrators and commissioners, was to "attract federal and private resources to aid in applying science and technology to urban problems."[24]

The new office attracted some federal funds. In fall 1972, money from NASA and the National Science Foundation enabled NASA and the city to collaborate on a one-year NASA/NYC Applications Project to explore technology transfer for city administration.[25] The money would pay salaries for two aerospace experts to work in city government. Despite past criticisms of "too many outside consultants," New York City administrators requested that NASA send two consultants to the city, one to be assigned to the Office of Science and Technology, the other to the Bureau of the Budget. (As a concession to critics, the city selected men with experience living in the New York City area.) Even before the funding was committed, city administrators contacted the American Institute of Aeronautics and Astronautics and several major aerospace corporations (among them, the locally based Grumman, which had played a central role in the Apollo Program), requesting names of potential candidates for the positions. (City files are filled with the resumés of applicants for these jobs).[26] Once hired, the two NASA consultants were charged with educating administrators about ongoing NASA projects and their potential for improving city operations. They were also to draft "problem statements" aimed at gaining better understanding of current needs. Specific projects included improving security in schools, graffiti prevention and removal, drug detection, and bridge inspection. Unsurprising, since the emergency services were the ones most analogous to defense and aerospace, several initiatives dealt with the police and fire departments. The new venture was short-lived, however: city finances were very tight, and since the program lacked continuing outside sponsorship, it concluded after only a year.

The record of aerospace management innovations brought to New York City is thus one of mixed success. In contrast to GIST's numerous applications, a lack of obvious results from the NASA/NYC Applications Project began to sour the enthusiasm for space age management. When Savas, the first deputy city administrator, left in 1972, he wrote in his letter of resignation to Lindsay that his half-decade of work had been "a kaleidoscopic combination of exhilaration and despair, rewards and disappointments, bitter frustrations and quiet satisfactions."[27] Such remarks expressed the feelings of many experts who believed that they had much to offer to American cities and yet could not seem to make their ideas work. City officials' continuing commitment to aerial photography thus contrasted sharply with their lack of interest in adopting satellite reconnaissance technology for comprehensive planning or as an input to GIST. Despite the presence of two NASA staff working on technology transfer in city government the year of the Landsat launch, interest in the satellite was nowhere to be found. Among the reasons that urban officials saw no need to alter their approach to planning after the development of GIST, two factors stand out: the compactness of the city and the ability, even from the earliest surveys, to see "Mrs. Smith's washing." Aerial photography and geographic information systems have remained essential tools for New York City administration to the present day.

New York City in Regional Context: Tri-State and LUNR

The move to integrate aerial photography and data from ground surveys for comprehensive planning was not limited to urban areas. Beginning in the late-1960s, renewed interest in regional planning was emerging across the nation. The New York City metropolitan area and New York State were early in the movement toward using geographic information systems as tools to address local planning in a regional context. Yet in both New York areas, city and state, NASA would be unable to convince system developers to incorporate satellite information.

The New York City metropolitan area, the intersection of three states, had a long history of regional planning. Organizations such as the private Regional Plan Association dated back to the 1920s. From its origins in the early-1960s, the Tri-State Transportation Commission (a public group that was renamed the Tri-State Planning Commission in 1971) began to investigate how it might create a geographic information system to synthesize various "reconnaissance" studies of the area with other land-use information. This system was developed throughout the decade, and news reports from 1972 describe how its successes stimulated NASA to suggest to Tri-State planners how satellite information might suit their needs. In a joint project with the Environmental Protection Agency beginning in April 1973, NASA astronauts took photographs of the metropolitan area from Spacelab and made them available for use.[28] Tri-State planners, however, never found applications for this satellite imagery.

The experiment that came closest to NASA's goal of making satellite reconnaissance part of comprehensive planning was LUNR (Land Use and Natural Resources Inventory), which was not a NASA-sponsored project. LUNR was an initiative of the New York State Office of Planning Coordination and Cornell University's Center for Aerial Photographic Research. In 1966, Nelson Rockefeller, the New York State governor, called for a statewide survey and inventory of land uses and natural resources. The plan was to combine information from aerial photographs with a variety of records based on ground surveys and thus give state planners a GIS. In these efforts to integrate aerial images into a computer system for statewide planning, GIS technology became applied to land-use issues of emerging national concern. By the late-1960s and into the 1970s, Congress was debating several versions of a national land-use bill that aimed at requiring localities to plan in line with regional goals and states to plan in line with national goals. The bill failed at the federal level, but many states passed laws making local development a regional matter.[29]

LUNR got off the ground in 1968, just as NASA's Earth Resources Survey Program was getting under way. That spring, there were survey flights over most of the state, creating fifteen thousand images in black and white. What was innovative about LUNR, what made it more sophisticated than New York City's GIST—and sparked interest in the project from NASA—was the project leaders' synthesis of aerial photographic data with a detailed array of information from maps and ground surveys. Accomplishing this goal required the

transformation of aerial photographic data into a form that could be introduced into a computer.

The New York State planning office contracted with the Cornell University
Center for Aerial Photographic Studies for much of the technical work, including photointerpretation. Cornell had a long history of work in photointerpretation. Photogrammetry courses and contract military research in the
College of Engineering dated from the 1940s, and back in the 1910s the university had hosted the army's aerial photography school. Led by Cornell's
Ronald Shelton and Ernest Hardy, the LUNR team conducted a pilot study of
LUNR in Cortland County to create a land-use classification system for coding
photographic information in numerical form. Standards for land-use
classification already existed, but the Cornell researchers chose not to adapt
LUNR to any existing standard (the use of aerial photography for land-use
classification dated back to the 1930s, when the U.S. Resettlement Administration's Land Classification Unit published a brochure explaining its uses).[30]
Rather, the team designed an idiosyncratic classification scheme based on in-
terviews with future users about their stated needs, a population of "numerous individuals and groups, primarily university and state agency personnel."[31] The aerial images, taken at a scale of 1:1000 and 1:2000, were reduced
and overlaid on USGS/New York State Department of Transportation maps of
a different scale, 1:24,000. Data interpreted from the photographs, combined
with other ground-based data, were then introduced into a computer system.

Since the Cornell team had little experience with computer mapping, they
contracted out some of the database development work to Carl Steinitz and
his colleagues and students at the Harvard Graduate School of Design's Laboratory for Computer Graphics and Spatial Analysis. Steinitz's colleague
Howard Fisher had created SYMAP, and Steinitz would write the software for
PLANMAP, for graphing and mapmaking, and DATALIST, for analyzing statistical relations among data sets, both of which were used in LUNR. At this
time, Steinitz and colleagues were also pursuing military contract research
(e.g., a pilot project for the U.S. Army Corps of Engineers, comparing land-use
classification systems). Like Emanuel Savas in New York City, Steinitz's work
drew from both defense and aerospace research traditions. At the design
school he offered graduate courses such as the 1968 "A Systems Analysis
Model of Urbanization and Change," taught with Peter Rogers in the Department of Landscape Architecture. In fall 1969, Steinitz, Richard Toth, David
Sinton, Frederick Smith, Douglas Way, and Timothy Murray taught a course

named "The Boston Region: Southeast Sector," which offered methods of landscape resource analysis combining airphoto interpretation with data processing technology to create a computer model of the region.[32]

LUNR became one of the most sophisticated early geographic information systems. It was able to map relationships between physical space and socioeconomic data and to display these relationships in map form. Sample maps in LUNR documents include, for example, state aid to towns per capita, percentage of families with incomes under $2,000, and percentage of housing units with sound plumbing. New data on land use, once entered into the computer, could be compared with older information such as aerial photographic records of New York State from 1938, 1951, 1955, 1958, and 1964.

By 1971, 275 data items already were in the inventory system, and there was room for up to 10,000. The total cost had run to a little more than $750,000, with inventory costs estimated at a comparatively modest $10 per square mile. Four computers were running LUNR data, and no special equipment was required. For these reasons, Shelton and Hardy expressed enthusiasm about its low cost, its large area of coverage, and the transferability of its techniques and technologies. They praised the system's simplicity, suggesting that the product could easily be exported for use in other states and regions by anyone who had the necessary computing power available.[33]

That year, the center began working with several local governments to help them make use of LUNR for area planning studies. By 1973, William Horne, assistant director of the New York State Office of Planning Services (the agency was renamed; it was formerly the Office of Planning Coordination), told an audience about how LUNR was being used for planning under HUD's section 701 policy. The system also was assisting in regional planning programs sponsored by both HUD and the state planning office. Businesses, too (e.g., Chase Manhattan Bank, New York Telephone, Sears Roebuck) had frequently used LUNR. Horne was enthusiastic: the technology was coordinating planning at the local and regional levels.[34]

Unsurprising, given that Governor Rockefeller's 1966 call to inventory the state specified the need for a survey of natural resources, the City of New York, the state's most built-up area, was not a major user of the system. Contracts with Lockwood, Kessler and Bartlett for aerial surveys of New York City and Long Island were the last to be signed. These urbanized areas, the last to be mapped, were overflown in 1969 and 1970, separately from the rest of the state. Further, while aerial photographs of the state initially were obtained at a scale of 1:1000 and 1:2000, they were reduced to 1:24,000 for standardization with USGS and state agency maps. This had practical consequences: the new scale was appropriate for natural resource studies and some business planning but not for physical and social planning at neighborhood level.

Efforts were made in the mid-1970s to update the resource inventory, some of the work being carried out as part of a pilot program called Land Related Information System (LRIS, 1974–77). Eventually LUNR was transferred from the state planning office to the New York Economic Development Board. State agencies occasionally updated data and provided users with information upon request, but there was no significant development of the system after 1978.

Satellite reconnaissance again was left out of the loop. State records in Albany offer evidence of discussions about possible use of satellite imagery, and Landsat images are held in state archives. The images are separated from LUNR files, however, and the satellite information was never coded and introduced into the LUNR database. Cornell's Hardy, like his colleagues at the Tri-State Regional Planning Commission, briefly investigated possible uses for Skylab imagery in related work, but did not follow through with planning applications. When LUNR developers presented their work at meetings on remote sensing, they characterized it as a form of "remote sensing." But in fact, the group never used imagery more sophisticated than black-and-white aerial photos. LUNR thus broke ground not in its use of newly available remote sensors but rather in its synthesis of statewide survey data, aerial and ground, into a GIS for comprehensive planning.[35]

Los Angeles

The first forty years of aerial photography in Los Angeles had much in common with the New York City experience. City of Los Angeles officials began using black-and-white aerial surveys in the 1920s, when Sherman Fairchild's company opened a branch office in the city. An LA–based community of aerial surveyors grew to serve the emerging market in the city and region (I. K. Curtis and F. M. Huddleston's Aerographic Company were part of this expansion). Beginning in 1927, Los Angeles County made use of federal and state mapping services for regional planning, and following World War II the job of aerial mapping was taken on by the California State Reconstruction and Reemployment Commission.

Between private surveys and publicly commissioned overflights for city, regional, and state agencies, Los Angeles was mapped and photographed repeatedly from the air, yet for decades most agencies did little resource sharing, a state of affairs much like New York City.[36] By the late 1960s, however, the paths of New York City and Los Angeles diverged. Local government in New York City, like most municipalities around the nation, continued to use black-and-white aerial photographs for physical and social planning, expanding its uses to become a data input to geographic information systems. Los Angeles officials tried a different experiment. Contacts with NASA-sponsored researchers catalyzed city administrators to investigate the uses of recently declassified remote sensors to assist in housing studies alongside more traditional land-use and comprehensive physical planning.[37] The city's Community Analysis Bureau, already experimenting with systems analysis and computer simulations in the context of community redevelopment, took on a new task: finding productive applications for color infrared data and GIS to achieve the overarching goal of reducing urban blight.

The larger story of space age innovations for city government in Los Angeles dates to 1962, when in the face of declining spending for California's aerospace industry, the state governor, Edmund "Pat" Brown, proactively set about to find new roles for industry executives and engineers. At the 1963 conference on Space, Science, and Urban Life, Oakland had been envisioned as the first laboratory for experimentation, but it was in Los Angeles that, with state funding, four of the earliest civil systems projects got under way. The city signed contracts with four in-state aerospace companies. North American Aviation studied transportation, Aerojet-General explored waste disposal and pollution, Space-General examined crime prevention, and Lockheed considered centralized information systems across municipal departments.[38]

Known as the "California experiments," these $100,000 contract projects (small for aerospace, large for cities) set the stage for similar experiments in urban centers across the nation. In related work several years later, for example, Caltech's Jet Propulsion Laboratory, one of NASA's main research centers, managed what became known as the Four Cities Project. With funding from NASA and NSF, four other California cities (Fresno, Anaheim, Pasadena, and San Jose) appointed technology consultants from local firms to investigate technology transfer to urban management with a focus on physical operations as opposed to social planning (the firms taking part were JRB Associates, a subsidiary of Science Applications, Inc.; Northrop Corporation; Space-General; and Lockheed Missiles and Space Company). Together these studies led by 1967 to what Ida Hoos estimated to be a multibillion-dollar endeavor, with forecasts for urban systems work by 1980 well over $200 billion dollars. Governor Brown, thrilled by the new direction for his state's massive aerospace industry, commented, "Can the kind of 'new dimension' thinking that found a way to get a Moon-probe off the launching pad also find a way to get able-bodied men off the welfare rolls? In California, we are finding out. And the preliminary answer is an emphatic YES."[39]

Despite such proclamations, the California collaborations, like the NASA/NYC Applications Project, proved to be short-lived. Beyond the preliminary analyses, the benefits for California's cities were not entirely clear. Even technological enthusiasts such as Robert Joyce, the Los Angeles CAB director, found a different answer to Governor Brown's rhetorical question about Moon-probe expertise: trying to solve "earthly problems," especially urban problems through aerospace innovations had shown that "transporting the astronauts from terra firma to land on the lunar sphere, travel hither and yon over its surface, and then back home to Houston" was a comparatively simple task.[40]

Lacking ways to quantify the benefits of these "experiments," lacking federal or state sponsorship to continue them, and asked by the state to prepare a comprehensive city plan, Los Angeles officials refocused their attention on aerospace innovations of a different sort: remote sensing technology and developments in photointerpretation. City administrators turned to the question of integrating aerial surveys into the city's developing information systems and using that data to improve housing and environmental planning. These efforts would be headquartered at CAB and in the Department of City Planning, the agencies already experimenting with cybernetic and computer simulation tools.

Remote Sensing Comes to CAB

The Community Analysis Bureau, the city's center for applying military management innovations to community redevelopment, was the first agency in Los Angeles to work with nonphotographic remote sensors. Interest in remote sensing at CAB followed directly from NASA-sponsored research. In the late-1960s, Eric Moore and his team of geographers arrived in Los Angeles from Northwestern University. Funded by NASA, their goal was to evaluate how remotely sensed information about housing quality might serve as a proxy for health indicators. Working with color infrared data that simulated the kinds of future information that would be available from Landsat, Moore reported that the pilot study was highly encouraging. Assessment of housing quality could not easily be made at the level of individual parcels, but at the level of blocks it was easier.[41] Testing remotely sensed data against other data from the LA County Public Health Department, Moore and his team reached an important conclusion—that the department could significantly reduce costs if they switched to his survey method.

County officials did not switch, but Moore's work grabbed the attention of CAB staff who since 1966 had been looking for ways to reduce urban blight through scientific and technical methods. CAB officials never committed to making use of satellite data for their community development work, the ultimate aim of NASA's sponsored research, but the Northwestern team's work with data from low-altitude aircraft led them to draw several conclusions. Robert Mullens, a CAB project analyst and remote sensing specialist, observed that color infrared aerial photography appeared to be "one of the most promising if not *the* most promising source of information on urban environmental conditions for urban analysis and planning."[42] Mullens already was familiar with photointerpretation. Earlier, while at the University of California-Riverside, he had been part of a NASA-funded research group attempting to characterize conditions in three neighborhoods in Los Angeles (East LA, Florence-Firestone, and Compton-Willowbrook—all low-income areas) based on aerial photography and comparing their photointerpretive findings with data collected on the ground.[43]

CAB leaders began to contemplate how the bureau might use recently declassified sensor technologies in housing surveys to suit its planning needs. In the same way that, earlier, staff had defined the city and CAB in cybernetic terms to facilitate the adoption of military innovations, now they drew close analogies between the bureau's mission to collect and analyze information about blight in order to slow its spread and what remote sensing could offer.[44] In 1970, shortly after Moore's visit, CAB staff began to experiment with photointerpretation. Borrowing black-and-white photos from Jene McKnight and colleagues at the LA Planning Commission and color infrared images from NASA aircraft studies (missions 56 and 73) that had been testing sensors for later use on Landsat, they consulted with outside experts (among them were Leonard Bowden, of UC-Riverside, whose research was sponsored by NASA and USGS, and M. C. Branch, of USC, who had just ended his nine-year term on the

LA Planning Commission). Adapting Moore's photointerpretation techniques to derive social data, CAB authors explored potential correlations among different information sources. A report from CAB that April presented the results of this initial study, including extensive discussion of what color infrared aerial images revealed about Los Angeles communities (fig. 8). CAB staff concluded that both aerial photographs and color infrared images had the potential to become important tools in assessing blight—and would be cheaper than other data gathering methods. From this early experiment, color infrared photography emerged as the standard for future CAB aerial survey research (aerial surveys did not, however, become standard for all planning activities in the city; other plans continued to use drawings).

In their examinations of borrowed images, CAB researchers quickly learned to value both advance planning and high-resolution images. They recognized that, rather than collecting large data sets for which an application later would have to be found (an approach that characterized Landsat), defining the problems they wished to study—in advance of the next round of aerial surveys—would make for less time spent looking at useless images. Scale, too, was a critical factor. Among the sets of preexisting photographs were images at scales from 1:6,000 to 1:12,000 to 1:24,000 to 1:50,000 to 1:60,000. CAB staff discovered that when resolution reached 1:24,000 (the scale of LUNR images) it produced "a definite loss of vital information if the topic under consideration is urban blight."[45] Thus, like planning departments in cities across the nation, CAB concluded that low-altitude aircraft would be the ideal survey platform for city and neighborhood housing planning. This decision effectively ruled out the future use of Landsat data, unless a new round of more sophisticated sensors were to be declassified.

The April 1970 report proposed a regular program of color infrared overflights to survey the city at 1:10,000 scale.[46] Lower-altitude flights were technically preferable, but resolution that was slightly less sharp was cheaper since pilots had to make fewer passes over the city. The plan was to conduct overflights annually—possibly twice a year, budget permitting. CAB staff would receive further training in photointerpretetation and learn how to compare aerial data with census information and other city records.

Thus, in March and April 1971 CAB commissioned an aerial survey of the city and surrounding areas—fifteen hundred photos in all, at a cost of $15,000. Flights at a scale of 1:10,000 and using color infrared sensors were flown during spring and fall 1971. Evaluations based on these aerial images

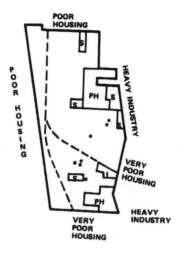

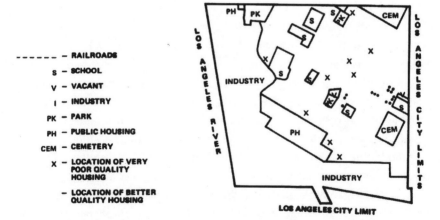

Fig. 8. Diagrammatic summary of findings from a study of housing quality in Watts, Los Angeles, based on an aerial survey using color infrared imagery. Los Angeles was one of the few cities to experiment with remote sensors of military origin soon after their declassification. Color infrared photography originated as "camouflage detection film." Maps from *A Practical Method for the Collection and Analysis of Housing and Urban Environment Data: An Application of Color Infrared Photography* (Los Angeles: Community Analysis Bureau, April 1970). Reprinted with permission of Los Angeles City Archives.

were then combined with housing inspection data from the city's Department of Buildings and Safety and 1970 census information "to produce estimates of unsound housing."[47] According to an October 1971 document, staff had difficulty developing correlations among the different data sources. In CAB reports (e.g., the October 1971 *State of the Tenth Council District*), color infrared images were used in the housing section with a caveat: exact correlations, between environmental conditions as shown by the aerial images and the socioeconomic information in the city's other databases, would have to await further study.[48]

Together, these information sources eventually became a valuable input to CAB's decision-making processes for combating urban blight. Information about unsound housing helped CAB staff identify the neighborhoods undergoing the most significant levels of change, which in turn helped the bureau in directing money for neighborhood revitalization. Analyst Mullens, in his 1972 presentation to the AIAA/PTI Urban Technology Conference, offered an enthusiastic assessment of how color infrared photography was aiding the bureau in several projects. From the outset, he said, a major benefit was cost savings: ground surveys of the same area would have cost an estimated $1,000,000.

At first, analysis of color infrared surveys proceeded separately from information system development based on military tools. But by 1972, CAB staff began to try to figure out how new aerial imagery might be systematically introduced into a database for more "scientific" decision making on housing policy. Across city agencies there was a desire to deepen the understanding of small areas and neighborhoods, and interest coalesced around a system that could integrate multiple data sources, map this information block by block, and model the outcomes for a variety of possible policies. Mullens described this information management tool as a "geo-reference system."[49]

CAB staff began work on such a system, which became known as the Urban Information System. Their reports characterize this comprehensive city planning tool as "a system to identify and remedy social, economical, and physical blight based primarily on aerospace experience," a system spanning not only the city but also the region, given that "problems of community blight usually extend across municipal boundaries."[50] A regional intelligence system for assessing urban blight, one that could be constantly updated, would be the ideal defensive weapon against the kind of neighborhood deterioration that led to urban unrest. It is notable that a chart pairing weapon system development flow and urban system development flow—the same chart that had appeared in other CAB documents on technology transfer—accompanied a a report on this system prepared by Robert Joyce for AIAA members. Perhaps because, on this occasion, the paper's target audience was the AIAA, not defense researchers, Joyce portrayed aerospace innovations as the roots of the decision-making system.[51]

Simultaneously, CAB staff set about developing another geographic information system, a small-scale system focused on housing analysis. Housing supply was an area of significant political debate in Los Angeles, as in many other U.S. cities. The Los Angeles Housing Model, as this project became known, took a systems approach, its goal being to formulate "equitable and financially reasonable" housing policies and programs, using the computer system to model and then decide among policy alternatives.[52] Removing at least some politics from the decision-making process would, it was hoped, speed the route to housing production across the city. The goal—to close the housing gap and estimate future housing needs—was in line with the 1939 Housing Quality Commission's aim to find housing for all Americans.

Work started on the model in 1972 and 1973. In order to introduce information from photographic surveys into a computer, administrators developed a mathematical scheme for coding housing and environmental quality. Like the field of land-use classification, efforts to numerically rate and mathematically analyze housing quality predated back the widespread use of electronic computers. In the 1940s, for example, the American Public Health Association had created a standard method for coding surveyed housing (the LA County Public Health Department used this system in simplified form). Earlier in the century, the U.S. Census Bureau had included questions about housing quality, but it discontinued this practice after 1960 due to concerns that the measures were unscientific.[53] Remote sensing enthusiasts suggested their method provided an antidote to the inaccuracies of other data sources, and (like the researchers on the LUNR project) CAB staff created their own classifications to turn color infrared aerial surveys into machine-readable form. Working with images from the 1971 overflights, staff coded housing quality on a scale of one to ten. Decisions were based on the assumption that differences in radiance (i.e., brightness in the form of pixel intensity) indicated differences in neighborhood conditions.[54]

The LA Housing Model was completed in 1974. It was not merely a model but a database with modeling capabilities. The project combined census data on housing, plus local housing inspection records, with aerial surveys. City officials could use the housing classifications and mathematical models developed at CAB to compare the predicted effects of alternative decisions on the perceived housing gap and to estimate future housing needs. Computer mapping technologies developed elsewhere (among them SYMAP and the U.S. Census Bureau's DIME, or Dual Independent Map Encoding) made it possible to display this information at the level of block and parcel. When urban managers were faced with alternatives in decision making, computer modeling could be used to predict and compare the various effects. Like LUNR, the system was created in a form that could easily be updated, but due to lack of funds new overflights were not completed. As late as 1976, the bureau was still using images from 1971.[55]

In their technical report, CAB staff characterized the new system as a parallel innovation to those being created by USAC, the federal interdepartmental committee sponsoring research on urban information systems. According to CAB's write-up, the system was "transferable in its present form," and in fact as soon as the Housing Model was created, data from it were requested by government organizations at local, regional, state, and federal level. HUD, which had partially funded the project, soon asked CAB to send along a copy of the software to Washington, D.C., for further dissemination. Locally, where the City Planning Department's Systems and Data Analysis Division had helped to integrate the aerial information into computer systems, staff used the Housing Model reports to develop community housing plans. City records do not, however, clearly link any specific policy decisions to the model. Nevertheless, the perceived transferability of CAB's geographic information system (again as with LUNR) was one reason that Robert Joyce, of the Community Analysis Bureau, William Horne, of the New York State Office of Planning Services, Donald Belcher, of the Center for Aerial Photographic Studies at Cornell, and Charles Robinove and James Anderson, of USGS, were among the handful of invited presenters at a 1973 workshop coorganized by Kodak (a contractor with NASA for Landsat sensors) and the Cornell University Center for Aerial Photographic Studies. The workshop's topic: Aerial Photography as a Planning Tool.[56]

LUMIS, City Planning, and the Jet Propulsion Lab

As CAB's experiments with color infrared images and a housing database and decision-making system took root, close colleagues at the Los Angeles

City Planning Department were organizing their own space age approach to citywide planning. The two city agencies already had been in joint discussions for several years about maximizing the uses of color infrared aerial photography. In addition to making use of CAB's imagery in the environmental and housing sections of its reports on community planning areas, the department entered into an independent partnership with the Jet Propulsion Laboratory (JPL), one of NASA's main research centers.[57] Their joint goal was to create a more sophisticated GIS to facilitate master planning at a new level.

The immediate impetus was straightforward: a project was under way to develop the area around the Santa Monica Mountains, and city planners decided that as their territory encroached on mountain terrain they needed to address issues of natural resource planning alongside their traditional focus on urban infrastructure. This transfer of aerospace technology to the urban realm was to include both management systems and remote sensing data. Funded by NASA, the collaboration between JPL and the city planning department led to the invention of LUMIS (Land Use Management Information System). In its pilot phase, LUMIS would use low-altitude aerial photographs and map only the LA City portion of the Santa Monica Mountains; after the pilot was completed, a regional planning system would be created. The regional system would use satellite imagery—an unsurprising choice, given that project reports were listed as NASA projects and that project leaders presented their early work at NASA's ERTS Symposium.[58]

The LUMIS development team included, among others, Los Angeles city planners Albert Landini and R. Wayne Bannister, the JPL's Nevin Bryant and Thomas Logan, of the Earth Resources Applications Group, and NASA official Charles Paul, of NASA headquarters. Project documents explain how, by the 1970s, staff at the JPL—like so many industry colleagues—were confident that their experience in the space program could directly apply to the new challenge of creating an urban information system. Expertise in information management had been one of the outcomes of research and development on flight projects such as Ranger, Surveyor, Mariner, and Pioneer, and JPL already had parlayed this expertise into work on LUNR and the Four Cities Program. Thus, LUMIS was merely another step along the path by which JPL hoped to "incorporate aerospace technology into problem solving mechanisms that directly affect the daily lives of average citizens," enabling its staff to work on problems in their own backyard.[59]

The project team began by surveying information needs and information resources available to city planners. JPL conducted a Delphi analysis of information needs for land-use planning in the city and compiled an encyclopedic record of overflights made by numerous contractors since 1950. These investigations confirmed that despite significant rhetorical support for comprehensive planning, in fact city agencies had shared little aerial survey data. In an effort to make comprehensive planning a reality, staff developed a method to combine land-use information from airphoto interpretation, digitized maps, and ground-based surveys. Aerial photointerpretation for the LUMIS project was contracted to the LUNR group at Cornell, who assigned numerical environmental ratings and land-use classifications to photos. This information was then translated into machine-readable form in order to offer land-use and natural resource information, census block by census block. A subsystem of LUMIS, LUPAMS (Land Use Planning and Management Subsystem) was created to offer parcel-level land-use information based on data on from the County Assessor's Office. LUMIS was innovative in that it merged many information files, including aerial images of many kinds and scales, maps of the Santa Monica Mountains area from 1970, 1972, and 1974, and details of 173 socioeconomic data items based on ground surveys. Like LUNR in New York State, LUMIS was a comprehensive GIS, able to combine natural resource data with urban socioeconomic indicators and display the data in map form.[60]

Project leaders at JPL did not stop there. The lab continued to tout its results and seek contracts with other cities and regions, and by the mid-1970s Pierce County and the City of Tacoma, Washington, were negotiating with JPL to create LUMIS systems. Optimistic that their system would be transferable to other users, the lab also pressed ahead with plans to expand LUMIS to include remote sensing data. A next-generation system—MILUS (Multiple Input Land Use System)—was scheduled for development starting in 1975.

The major innovation planned for MILUS was to be the addition of satellite information (fig. 9), but this follow-up system, like other satellite-input GIS systems such as ERISTAR (Earth Resources Information Storage, Transmission, Analysis and Retrieval), was never funded. As NASA officials were discovering, city governments had little use for satellite information. Records in the LA City Archives suggest that city officials instead turned to integrating the multiplicity of information systems in use throughout the city. A March 1975 letter from Mayor Tom Bradley to the city council updated them on ongoing ef-

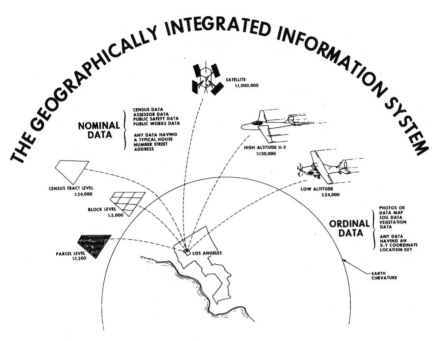

Fig. 9. Scale was a critical factor in using aerial surveys. This Jet Propulsion Laboratory sketch shows various survey methods and the scale of data produced by each. The JPL, here illustrating its design of a GIS for LA in 1976, planned to integrate ground data with aerial imaging and satellite-derived information. The city continued to use aerial surveys, but no system using satellite data was ever built. Reprinted courtesy of NASA/JPL/Caltech.

forts to plan a citywide management information system to merge records across departments. By executive directive, Bradley ordered city administrators to "lay the groundwork for the development of an integrated information systems plan."[61]

Evaluating the Results

A decade after William Mitchell had expressed concern about technology transfer from aerospace to urban operations, meetings on "Aerospace Technology Transfer to the Public Sector" were still in full swing. Yet their tone had shifted. At a 1979 conference cosponsored by NASA, AIAA, and PTI,

Cincinnati's city manager, William Donaldson, expressed now-common frustrations with the much heralded but little proven promise of technology transfer from aerospace to urban needs: "I see many people who might otherwise be on the unemployment rolls engaged in organizing groups to promote technology transfer." And yet, he added, "I still have the same old garbage truck, the same street sweeper that doesn't work, and in general the same system that does a very poor job of delivering the rather simple kind of services the people I work for depend on."[62] Donaldson pointed to the powerful network of individuals and institutions who had committed themselves to technology transfer and suggested that this was the primary reason such efforts had persisted, for years. When aerospace innovations became the focus of technology transfer, many special interests were satisfied, but urban operations were rarely improved.

The experiences of New York City and Los Angeles reveal the multiplicity of ways in which the space age made its way into city management. City administrators, while importing tools from the defense research community, such as systems analysis and computer simulations for a more scientific approach to community renewal planning, were, toward complementary ends, also cultivating ties with the aerospace industry and NASA. Whether the goal was urban redevelopment, continuous master planning, or housing evaluation, for a time the importation of new management methods from defense and aerospace seemed to promise a cure for chaos.

Observations by early critics, however, soon grew to a chorus excoriating the space age approach. Urban professionals expressed frustration that their work had been hijacked by refugees from aerospace, arguing, like the author of an editorial in *American City*, that public gratitude should lie with the long-suffering civil servants and urban planners, not the double-talking systems experts who understood little about the nature of urban problems. Even some members of the aerospace community began to criticize their own efforts publicly. McDonnell Douglas's Arnold F. Goodman, for example, suggested that scientists and engineers with solutions and techniques "in their pockets" seemed to walk around in search of problems that fit. There was, he wrote, an important distinction between "solving mathematical problems" and "solving problems with mathematics." He chastised his colleagues for overlooking that distinction.[63]

Among the most vocal critics was Ida Hoos, a researcher at UC-Berkeley's Space Sciences Lab. In her critiques of military-to-urban technology transfer,

Hoos returned to the source of enthusiasm for RAND's analytic innovations and showed how their original military applications were not as successful as they initially appeared. Her criticisms of space age management followed a similar line of argument. The lack of obvious positive results from space age management in U.S. cities was unsurprising given that systems "cited as exemplars" of successful space age management were in fact "prime examples of miscalculation and mismanagement" in her estimation; she pointed to cost and time overruns, and inconsistent results, as par for the course. SAGE, she said, "obsolete before it was completed and long before it was paid for . . . was successful only because our enemies failed to attack."[64]

Once again, a culture clash impeded technology transfer. This time the clash had little to do with the hierarchical and centralized command-and-control organization of the military versus city governments; it had much to do with differences in money and technological infrastructure. Large-scale federal investments in research and development for defense and aerospace—even in periods of spending cutbacks—significantly outstripped resources available to urban areas. Since a technology's effects vary depending upon the context in which it is applied, urban planners and managers lacking the same financial resources were unable fully to adapt military and aerospace innovations to the problems they believed required attention—except rhetorically. Lacking the resources necessary to develop in-house analytical and technical expertise, city administrators largely were dependent on outside consultants from companies such as McDonnell Douglas. The companies' excellence in supplying products and services to military and intelligence agency clients did not translate easily into adapting aerospace innovations to the urban context. Harry Finger and John Lindsay tried to replicate the DoD–NASA model—to institutionalize staffs of technical experts who could independently review, intercede, and redirect efforts as necessary in work that was contracted out—but neither HUD nor New York City had the resources to augment agency staffs for the long term. Despite earlier claims about the similar challenges faced by aerospace and urban managers, and despite repeated efforts to collaborate on "problem definition," enthusiasts in both communities began to acknowledge the chasm that divided them and the difficulty of "building bridges between the aerospace scientists or department technician on the one hand and the city department manager and his professional staff on the other."[65]

As Mark Keane, executive director of the International City Management Association explained, scientists and engineers had been invited to work in new laboratories—American cities. Yet obstacles, from skeptical "laboratory administrators" to a continued lack of financial resources, were preventing the nation's scientific and technical experts from doing their best work.[66] For a decade, Keane and his colleagues remained optimistic. Leaders from aerospace, like their close cousins in defense research institutions, acknowledged early difficulties and yet continued confidently to place their bets on the value of space age innovations for American cities. It was not until the late-1970s that these technological enthusiasts recognized that in some areas the two cultures could not be bridged.

Like the "systems approach," a phrase whose vagueness made it an ideal shared rallying cry for both defense intellectuals and city administrators, the notion of "space-age management"—which even NASA's Webb had described as difficult to define—would prove to be largely a rhetorical ploy, an opportunity to fasten the ideal of technocratic rationality onto the messiness of city management.[67] After a decade in which results for cities from partnerships with aerospace were not observable on the ground, space age management began to fall out of fashion. USA, Inc., unable to obtain a single contract, folded in 1969. HUD, which had committed itself to its own space age housing development program in 1969, Operation Breakthrough, abandoned the program by the mid-1970s.[68] The AIAA discontinued its meetings with PTI and NASA on technology transfer after 1979.

The fate of satellite use in city government was equally bleak. Although well-suited to some civilian applications, such as agricultural studies of crop inventories and deforestation monitoring, uses for satellites in the urban context, particularly in urban social planning, were far more limited. An annotated bibliography on remote sensing for planners compiled by the National Technical Information Service (NTIS) reflected this fact. Although the bibliography's title referred to urban planning, the articles it listed were primarily concerned with natural resource planning; only a few dealt with urban planning projects (most of them large-scale).[69]

The failure of Landsat to become an integral component of city planning and management has many explanations. One relates to NASA's approach to promoting data use. Data acquisition, and learning from that data over time, ideally should be cumulative. Yet much of NASA's sponsored remote sensing research did not apply findings from early studies to new areas. Instead, researchers repeatedly focused on experimental validation, testing and retesting different types of remote sensing observations and comparing them to ground surveys. This phenomenon was in direct contrast to USGS efforts to create a standardized land-use classification system that could synthesize remote sensing with ground-based survey data. The NASA-funded experiments validated the reliability of remote sensing techniques in a variety of settings and helped to steer a course for academic geography, but they did not go a long way toward incorporating remote sensing into day-to-day practices in city planning agencies. This was compounded by the fact that far more academics than local government officials participated in these experiments, providing applications of satellite data that were almost always a step removed from urban managers' needs. NASA, criticized elsewhere for confusing technology transfer with information dissemination, was culpable in this area, too.[70]

The disappointing results were also due to the imagery's level of resolution. Landsat's early flights simply did not produce images that were sharp enough for neighborhood and housing planning applications. For example, in a 1973 presentation, USGS's Robinove noted that the standard resolution of Landsat data was 1:1,000,000, with enlargement possible to 1:250,000 and in some cases even 1:100,000.[71] This was not adequate for physical and social planning, where since the 1920s aerial surveyors consistently had agreed that the scale generally had to be less than or equal to 1:2,400, or 1 inch to 400 feet, and resolution far sharper would be needed in maps prepared for tax equalization, factory layout, and planning of parks, playgrounds, cemeteries, and golf courses.[72] Unfortunately for NASA, resolution would not change substantially until the 1990s, long after Landsat had been privatized. While higher resolution was technically feasible before this date, and was being used in military reconnaissance and by USGS mappers in its classified lab, the space agency chose not to take a possible security risk by disseminating high-resolution images.[73]

The most compelling explanation for local governments' lack of interest in Landsat data is simply that use of the alternative, aerial photography, was widespread and that aerial photography was not among the urban management practices that were blamed for the urban crisis. It became obvious that when comparisons were made between satellite imagery from Gemini and Apollo, high-altitude imagery from aircraft-based remote sensors, low-altitude aerial photos, and ground-based survey data, certain types of information about urban areas were not visible from declassified civilian spacecraft platforms. For example, in a NASA study of Houston, investigators mapped the area using Landsat, using low-altitude aerial photographs as a source to verify satellite-based land-use classifications. They concluded that the accuracy of urban classification from satellite data was only 40 to 70 percent, compared with 70 to 90 percent for rural areas.[74] Part of the difference was resolution, but the other explanation was that with nonphotographic satellite technology, investigators needed training to interpret even basic classifications from the spectral details of the information received. Without the kind of large-scale funds of the military space program, city administrators faced an additional barrier to committing to the technology, even if it seemed to make their jobs easier by capturing images in GIS-compatible digital form. Once users developed scoring methods to make it possible to introduce aerial photography into GIS, the fate of satellite surveys was sealed.

Unwittingly, then, the technology that appeared to lay the groundwork for satellite surveys in fact appears to have impeded their dissemination. As it used low-altitude aerial photography to validate the satellites it wished to popularize, NASA helped to promote the use of aerial photography and low-altitude aircraft-based remote sensing. The space agency continued to fly aircraft missions in its High Altitude Aircraft Program, but private survey companies flying low-altitude missions dominated city planning markets. Their black-and-white aerial surveys, used alone or integrated into GIS, continue to be used in cities today.[75]

Enthusiasm for the space age within American cities may have been short-lived, but the aerospace community's mobilization for urban problem solving had long-lasting, if unpredicted, effects. Like their colleagues at RAND, disappointments in a specific endeavor did not mean an end to the continuing push for market expansion. Industry giants (e.g., Lockheed Martin and Northrop Grumman) have maintained civil systems divisions for nonmilitary clients. The JPL has created a Cartographic Applications Group to direct GIS development for both military and civilian clients. NASA has continued to fund remote sensing research through its Earth Science Applications Division and the NASA Stennis Space Center Commercial Remote Sensing Program. The Universities Space Research Association, established in 1969, has served as the successor to the Sustaining University Program, nurturing ties between academic researchers and the civilian space program. PTI has helped to sustain relationships between technology manufacturers and local governments. And LUPAMS, significantly updated, has continued to serve the City of Los Angeles.[76]

Like their colleagues at defense research institutions adapting command, control, and computing innovations to the urban context, the aerospace community's efforts to bring intelligence, surveillance, and reconnaissance developments from the space program to cities emphasized a top-down approach to planning and management. Such approaches were not without controversy in an era when protests were mounting to change the power relationships between city residents and their government. As defense and aerospace experts eagerly applied innovations to centralize city planning and management, many voices were calling for community participation and decentralization in urban decision making, approaches already formalized in the Community Renewal Program and the Economic Opportunity Act.

An alleged motivation for adopting systemic approaches to planning and management was the cybernetic emphasis on "feedback." James Webb had been insistent that space age urban management be open to vigorous citizen participation and dissent.[77] Yet when these management methods were introduced, urban activists saw little evidence that their views had become part of the system. As one observer explained, many "black power leaders" were adamant that space age techniques and technologies remained "instruments of oppression" because they did not accommodate adequate community input or debate.[78]

Webb could lecture mayors about how the space program's innovations would meet their population's needs, but results were difficult to see. Despite attempts to convince city leaders that spin-offs from defense and aerospace would trickle down to improve the lives of citizens directly, grassroots resistance was significant. When the Apollo-capsule disaster killed three astronauts in 1967, the reaction in some quarters was cool. "Black people did not join the nation in mourning," wrote Julius Lester, because the space program had diverted funds from city problems. "White folks trying to get to the moon, 'cause it's there. Poverty's here! Now get to that!"[79]

These criticisms would not go unheard. At the same time that they were applying techniques and technologies of command, control, computers, intelligence, surveillance, and reconnaissance to plan and manage cities, the community of defense intellectuals began to cultivate relationships with urban activists, seeking ways to improve communication between city leadership and the grassroots. In these outreach efforts, their strategies for urban problem solving grew to embrace community participation and decentralization. At first glance, these approaches appear to stand in direct contradiction, but as the next two chapters document, this was not the case.

Part III / The Urban Crisis as National Security Crisis

CHAPTER SIX

Cable as a Cold War Technology

RAND Corporation analyst Paul Baran was keenly aware, in the early-1960s, of the importance of decentralizing U.S. infrastructure to prepare for a nuclear attack. Working on contract research for the U.S. military, Baran pointed to the vulnerabilities of the nation's centralized defense communications systems. A single strike could disable the entire network. Baran sketched plans for a distributed system designed to survive a nuclear attack, and ARPANET, the military's precursor to the internet, followed from his proposal.[1]

With his RAND colleagues seeking new contracts for domestic urban research, Baran considered how his ideas about defense communications might have application in American cities. Together with an MIT professor of management, Martin Greenberger, who also at that time was working on defense-sponsored communications research, he proposed that distributed communications "help alleviate some of the urban sores which previous technologies have aggravated."[2] Baran and Greenberger's 1967 paper *Urban Node in the Information Network* characterized U.S. cities as overgrown "nodes" in the nation's information infrastructure. They urged that steps be taken to disperse urban populations and use cable-based communications to maintain

human connections. While their call to disperse urban populations stood outside the mainstream in 1967, Baran and Greenberger were not alone in seeing
promise for attacking the nation's urban problems by using cable communications. A cadre of defense communications experts shared that view.

Cable is a television broadcast system that delivers information by underground coaxial cables rather than via over-the-air signals. A popular entertainment medium today, cable has a history that is not so well known—a history
that is intertwined with the public fears of domestic social unrest that grew
from deteriorating race relations in the nation's urban areas. Cable was not of
military origins, but in the hands of defense intellectuals in a period of national conflict, it was envisioned as a tool for maintaining domestic stability
and order. In this vision, the medium could end the alienation of the "ghetto
dwellers" who were believed to have precipitated the violence in city streets.
This alienation would end, not through the broadcast of psychological warfare
but through the improved delivery of social welfare. These experts imagined
services piped into every house and apartment via a municipal cable network—from banking, shopping, and adult education to medical consultations,
community-produced programs, and town meetings.

Histories of American science and social science have described how, during the cold war, national security priorities shaped the trajectory of many academic disciplines, from physics to psychology. Communication research, in
both theory and application, also was influenced by such priorities. Scholars
have documented many uses of media, including radio and television, to disseminate American propaganda and psychological warfare both abroad and at
home, using militaristic, hierarchical approaches to communication.[3] Yet military planners and managers embraced strategies of both centralized and decentralized control. During the cold war, an expanding definition of national
security strategy grew to encompass economic and social development operations overseas alongside traditional combat. The invention of "interdisciplinary war" laid the social science foundations for a set of defensive strategies
that could be interpreted as the antithesis of military operations. Ideas about
development were chief among them, balancing political action and political
stability, simultaneously to satisfy citizens' demands for community participation and political leaders' calls for social control. Similar strategies would be
applied to urban development efforts in the War on Poverty at home.

The reappropriation of these strategies, embedded in many of the era's social policies and programs, can be made visible through an analysis of the de-

fense intellectuals' plans for community development through urban cable communications. Like the military network that Baran envisioned, providing security through decentralized communication, during the years of urban crisis, alongside their efforts to export defense and aerospace innovations to centralize administrative decision making, many defense intellectuals saw an essential complement when they looked at decentralized citizen-produced cable programming. Cable became a "cold war technology" during its first decade in American urban centers—1966 to 1976—as its framers envisioned uses in line with community development programs of the period, both overseas and at home. Understood in this context, the landmark 1972 decision by the Federal Communications Commission (FCC) to mandate channels for citizen programming and municipal information on urban cable systems can be reinterpreted as an artifact of the widespread belief that these uses for the medium would maintain domestic security by reducing the citizen alienation that precipitated urban violence.

Cold War Social Science

The importation of social scientists and their methods to military and intelligence work did not begin in the cold war. Applications of social science research can be traced to operations during World War II and before. Yet it was at the height of the cold war, during the 1950s and early-1960s, that specialists from the defense and intelligence communities worked increasingly closely with universities and think tanks to create social sciences for national service. Social scientists across an array of disciplines were recruited to join defense experts and systems analysts, becoming partners in achieving foreign policy goals increasingly characterized in social welfare terms.

Adam Yarmolinsky, Christopher Lasch, and others have documented how the institutional alliances of this period were the result of deliberate agenda setting at the highest levels of U.S. military and intelligence hierarchies. These alliances were based on an expanded definition of national security strategy in U.S. foreign policy—that stability in overseas regions needed to encompass actions designed to reach civilian as well as military populations. From the 1950s, military and intelligence agencies, in partnership with foundations such as Ford and Rockefeller, began to seed the creation of new research centers at elite institutions to fill in the outlines of this expanded definition. At RAND, Harvard, MIT, and Columbia, to name just a few, centers for communication research, area analysis, development studies, and behavioral sciences created interdisciplinary communities of scholars whose research was explicitly or implicitly tied to security concerns overseas.[4]

At RAND, for example, researchers began to incorporate socioeconomic analyses of overseas populations into regional analyses for their military sponsors. RAND stepped up its hiring of social scientists, and the social scientists' work with systems analysis in turn reshaped the "systems approach."[5] RAND's own story mirrors larger transformations in both American social science and U.S. military strategy during the 1950s and 1960s. Just as information system developers for the Department of Defense were incorporating socioeconomic information into command-and-control information systems (e.g., the Hamlet Evaluation System) and as airphoto interpreters were beginning to deduce socioeconomic details about populations from reconnaissance imagery, security analysts at RAND turned their attention to socioeconomic data in their studies of political systems.

The Center for International Studies at MIT (CENIS), funded by the Ford Foundation as a front for CIA-sponsored research, served as a headquarters of this new social science. Established in the 1950s, CENIS's interdisciplinary community of scholars laid the foundations for the field that became known as development theory. Daniel Lerner, Lucian Pye, Ithiel de Sola Pool, Walt Rostow, Guy Pauker, Suzanne Keller, and Raymond Bauer were among the affiliates for whom basic social science research in fact was often closely tied to military and intelligence agency needs. Their collective creation of a set of ideas about stages of development was intended to provide sponsors with blueprints for controlling modernization overseas to fight Communism. (Michael Latham reports that several of President Kennedy's advisers, including McGeorge Bundy and Walt Rostow, supported the creation of a "modernization institute" for the State Department and the DoD, but they discarded this idea in favor of sponsored academic research, like the ongoing work at CENIS.)[6]

Command, control, computers, intelligence, surveillance, and reconnaissance were all essential components in military planners' decision-making arsenals. The goal of social engineering for the long term demanded a different approach, however—that the populations whose development was being engineered take an active interest in the process, rather than be coerced. The development theorists identified citizen participation and communication technologies, from telephones to mass media, as essential means to serve this end.

Community participation would invest populations in their own moderniza-tion, and media would provide informational support. According to this view, communication—among citizens and between citizens and government—was an essential component of controlled development. As it turned out, these ideas could easily be integrated into the cybernetic worldview wherein "feed-back" was essential to achieving the goal of "homeostasis"—stability and se-curity—in a political system.

Development theory was first applied in U.S.-sponsored programs in the Third World, and later during the Vietnam War.[7] From its roots in the Ken-nedy administration, the war in Vietnam offered opportunities for social sci-entists to apply their theories about citizen participation, communications, and development to a real conflict. Prominent among them was a "pacificat-ion program," whose goal, according to Colonel Erwin Brigham, chief of the Research and Analysis Division, Civil Operations and Revolutionary Develop-ment Support at the Military Assistance Command Headquarters in Vietnam, was reducing counterinsurgency. In Brigham's words, pacification was "estab-lishing or re-establishing local government responsive to and involving the participation of the people."[8] Collaborations between civilian intellectuals and military officers brought social science theories to bear on a range of re-lated foreign operations. Amrom Katz, a military reconnaissance expert at RAND and later assistant director of the U.S. Arms Control and Disarmament Agency, called it "interdisciplinary war." "Wars," he said, "are getting less mil-itary."[9]

Enthusiasm for interdisciplinary war was reaching its peak in the mid-1960s. In 1967, the *Report of the Panel on Defense Social and Behavioral Sci-ences,* issued by the Defense Science Board of the National Academy of Sci-ences, confirmed that the defense establishment perceived success in its pa-tronage. Whether for traditional military operations or for peacefare activities, the U.S. military had managed to enlist an "eminent group" of social scientists in research for national service.[10] Several members of this group were invited to play more formal roles on the national political stage. CENIS's Walt Rostow, for example, became an adviser to Presidents Kennedy and Johnson, first as chairman of the State Department Policy Planning Council and later as special assistant for national security affairs.

In this climate of confidence, it was not long before the intellectual allies of the defense establishment began to speculate that, as with the systems analy-sis and aerial reconnaissance they originally had developed for military operations, much internationally focused social science research, particularly tech-niques derived from studies of communications and development, might serve in softer defense operations at home. Like his colleague Paul Baran, RAND's president, Henry Rowen, endorsed the idea. In a 1968 speech to the Twenty-second Military Operations Research Symposium at the U.S. Naval Postgraduate School, Rowen explained the close connections he saw between the nation's traditional military concerns and domestic issues that on first glance appeared to bear little relation. "The problems of national security and national welfare are not neatly separable from either a research or a policy standpoint; they form part of a continuum," he asserted, suggesting that many of the issues faced in foreign policy had analogs on the home front. These ranged from the violence motivated by social inequalities to the extent to which suppressive measures could be taken in a given region.[11] Despite the limitations of other defense and aerospace innovations in city governments already emerging at this time, defense intellectuals such as Rowen remained assured that their tactics were transferable. By conceptualizing the urban crisis as a national security crisis, parallels between external and internal threats, be-tween insurgencies overseas and those at home, could easily be found.

The nation's political leadership had for some time already expressed its fears of internal threats, both publicly and through the use of covert opera-tions. Annual hearings in the Senate on the state of the nation's domestic se-curity were made possible by the McCarran Act—the Internal Security Act (1950). The FBI, under the long reign of director J. Edgar Hoover, intensified its close watch of individuals and organizations. For example, the FBI had monitored the National Association for the Advancement of Colored People (NAACP) since 1941 (action initiated by Roosevelt), and in the 1960s the bu-reau stepped up its attention to other "subversive" participants in the move-ment for civil rights. U.S. Army domestic intelligence operations surveilled groups advocating social change (e.g., the National Urban League) and opened files on a hundred thousand Americans (among them, Martin Luther King Jr.). The Central Intelligence Agency (CIA) had its own program, Opera-tion Chaos, which monitored dissident activity in the United States between 1967 and 1973. The concept of "black terrorism" dates to this period.[12]

The central questions for these domestic military and intelligence opera-tions included: To what extent were the black power, civil rights, and other dissident groups allied with Communist causes? More broadly, to what extent were urban leaders facing an internal threat? With race relations increasingly defined as an issue of national security, economic development and social welfare for black Americans and more contracts with the defense intellectuals became a civil defense priority. Soon these men came to believe that their interdisciplinary strategies to control violence and to engineer development outside America's borders equally could be applied to maintain social stability at home. With active support from the Johnson administration, their approach became official policy.

Thus, if during the 1950s and early-1960s U.S. foreign policy goals increasingly were characterized in social welfare terms, by the late-1960s the nation's domestic social welfare goals increasingly were characterized in military terms. The Johnson administration's War on Poverty and War on Crime expressed this tone in urban and social welfare policy. So, too, did its staffing of economic development programs. For example, Adam Yarmolinsky moved from his position as special assistant to the secretary of defense to become deputy director of the president's Task Force on Poverty, helping to administer the Community Action Program. Joseph Califano, assistant to McNamara in the Office of the Secretary of Defense, became President Johnson's special assistant for domestic affairs. William Gorham, deputy assistant secretary of defense, became assistant secretary for planning and evaluation at the Department of Health, Education, and Welfare. In this context, Great Society initiatives, widely remembered as a multifaceted response to poverty, educational inequalities, and hunger, can be reinterpreted as measures to provide controlled development to bolster the nation's internal security from Communism and from further violence.[13]

At the beginning of the cold war in 1946, Tracy Augur had compared the urgency of civil defense with that of urban planning and concluded, pessimistically, that "Institutes for Research in Urbanism" would likely never be built on a scale like those of defense research institutions. Just two decades later, the context had changed. Defense interests and urban policies became aligned. Think tanks—among them RAND, MITRE, SDC, and IDA (the Institute for Defense Analyses)—formerly exclusively committed to defense contracts, including overseas projects, created new urban systems divisions, committing their attention and methods to problem solving for cities. Two think tanks in particular embody this alignment: the Urban Institute, which opened in Washington, D.C., in 1968, and RAND, which opened its outpost in New York City the following year. These institutions, staffed with researchers schooled in military systems analysis and, increasingly, the social sciences, mobilized for urban experiments based on citizen participation. These intellectual allies of the military establishment would play significant roles shaping domestic cable policy in the image of their expanding conception of national security.

In 1969, H. L. Nieburg, a political scientist and scholar of political violence, wrote about the growth of a new "urban problems industry," composed of commissions and think tanks. He cynically noted how "just as every energetic pressure group capitalized on the cold war . . . so the explosion of black ghettoes . . . furnish a new wave of special-interest slogans."[14] What is so remarkable in the story of American urban development is how an expanding definition of national security offered the cold warriors new opportunities for work in these ghettoes. Equally remarkable is how they—and many other constituencies, in turn—came to believe that a new generation of communications technologies, thoughtfully applied, would directly attack the nation's urban problems.

Linking Communications and Urban Problems

For the new breed of urban expert who emerged in the late-1950s, the defense intellectual, the prevailing image of the city was a communication system. These men defined their work in concert with new technologies: cities became cybernetic systems, and computers became potential tools to improve the quality of urban management.

The analogy between cities and communication systems was extended in the late-1960s. If cities were communication systems, these experts proposed, then by extension urban problems were communication problems. RAND analysts Herbert S. Dordick, Leonard Chesler, Sidney Firstman, and Rudy Bretz observed that exchange of community information was sadly lacking in America's inner cities. In a report that circulated both as a RAND paper and as part of a staff paper for President Johnson's Task Force on Communications Policy, they also pointed out that communication between the inner cities and other city neighborhoods was nearly nonexistent. The "problem of communications" was hampering ghetto residents from entering urban job markets and even from gaining access to the social programs specifically established to assist them.[15] "Deterioration and distortion of the communications system," they diagnosed, had become a problem requiring immediate attention.[16]

Analysts seeking explanations for the failure of urban renewal, the nation's prior large-scale approach to attacking urban problems, found many. Among the most prominent was that renewal, with its emphasis on physical infrastructure planning, did not give enough attention to social welfare planning, and that in particular the program lacked concern for citizen participation. Following directly from the social science research undertaken for application in foreign operations, researchers increasingly framed the central challenge of ghetto life as isolation. They concluded that many of the problems faced by inner-city residents were due to failures in communications. Urban sociologists (for example, Louis Wirth and Michael Harrington) had long characterized America's poor as isolated from the rest of society; the research of the 1960s and 1970s added an emphasis on communication. A variety of social science experts came to support this view, their fields ranging from urban planning, with its cybernetic conception of city processes, to urban sociology, which developed a network model of social interactions, to communication research, where scholars turned their attention to questions about how inner-city residents obtained information and to what extent mass media played a role. The Kerner Commission report on civil disorders concurred. It identified the homogeneity of news media and the lack of outlets for citizens to air their views as critical components in the disconnect between black and white America, between minority citizens and government.[17]

In the context of urban crisis, it became a national priority to find ways to reduce citizen alienation and to encourage nonviolent participation by racial minorities in city planning and management processes. The Economic Opportunity Act was an early example of a policy in this spirit: it stipulated that antipoverty programs had to be administered with the "maximum feasible participation" of community residents. If, as so many social science experts proposed, communication and community development were linked, the next generation of urban social welfare policies would have to attend to urban communications infrastructure.

Discussions during the late-1960s and early-1970s that linked communication technologies to community development and urban social welfare engaged a diverse group—defense intellectuals, grassroots activists, politicians, and academics. Some—for example, the defense intellectuals and media activist Paul Ryan (Marshall McLuhan's research assistant, who coined the term "cybernetic guerilla warfare" to refer to activist television)—explicitly grounded their claims in cybernetic thinking. Whatever the philosophical roots of their arguments, a multiplicity of individuals and institutions—RAND, Raindance (a counterculture think tank), the National Technical Information Service in the Department Of Commerce, the Urban Communications Group, the Urban Institute, the ACLU, the National Academy of Engineering, and the Electronic Industries Association—came to express the view that the problems of cities were communication problems.[18] It followed that communication technologies, thoughtfully applied at both the neighborhood level and city-wide, might offer remedies.

The burgeoning interest in the cable medium that many defense intellectuals expressed during the late-1960s signaled an important expansion of their understanding of successful urban development in the years following widespread urban unrest. A growing concern for community participation reflected an evolution in their assumptions about how best to address urban problems.[19] Centralized, hierarchical strategies, well-suited to military decision making and psychological warfare, would have to make room for a new emphasis on public participation. Cybernetic definitions of city systems had emphasized "feedback," but early efforts to gather this feedback had been limited to "quantifiable and depersonalized" data.[20] As researcher Joel Edelman wrote in a memo to all personnel at the New York City RAND Institute, studies to date had focused on analyzing city agency performance on urban problem solving at the expense of understanding public perceptions of city agencies and urban problems.

Seymour Schwartz encapsulated the rationale behind this new emphasis at an Association for Computing Machinery meeting ("The Application of Computers to the Problems of Urban Society"). Schwartz cited several studies that highlighted the alienation of individuals in mass society (among them, Kenneth Keniston's *Young Radicals* [1968] and Theodore Roszak's *The Making of a Counterculture* [1969]), and he urged his technocrat colleagues to reconsider the consequences of rational, hierarchical, closed-door decision strategies in favor of processes that invited public participation. This "more democratic process" might "produce worse results," but it would respond to the increasing sense of alienation among the nation's urban population.[21] Schwartz's remarks point to the growing realization that the technical and technological decision-making tools used in the nation's Community Renewal Program and other comprehensive planning efforts were insufficient improvements to decision making. These tools had failed to provide citizens with the outlets for participation that were increasingly recognized as fundamental to democracy and to reducing the sense of alienation that the experts believed had precipitated urban violence.

By the latter half of the 1960s, the enthusiasm for "objective" decision-making tools that characterized efforts to transfer defense and aerospace techniques and technologies to city administration was being supplemented by a more nuanced sense of the need to strike a balance between "objective" outcome and participatory process, between managerial oversight and community control. Command, control, computers, intelligence, surveillance, and reconnaissance would need to embrace communication. As the development theorists had outlined, two specific tasks had to be accomplished: improve communication among citizens, especially black and white, and improve communication between citizens and government. The most likely candidate for this job was cable.

Cable and the City: A Brief History

Historians have written little on the interactions between debates about urban problem solving and debates about cable television policy. Yet the two coincided. It was during the transformation of cable from a rural to an urban technology that cable regulation became an issue of national communications policy. The first cable systems were developed in the 1940s as technical solutions to retransmit broadcast signals for residents of mountainous areas with poor television reception. Beginning in 1959, when it declared cable to be a local phenomenon, through 1965, the FCC chose not to regulate cable. This absence of regulation likely stimulated the growth of cable systems, which during that period were alternatively referred to as community antenna television (CATV), community television, cable communications, and broadband communications.

The medium we now call cable television did not share the spectrum limitations of its broadcast television counterpart, and by the mid-1960s cable's increasing penetration into urban areas offered glimpses of a possible alternative to the limited programming, and limited channels, of broadcast television. As was the case during the early history of radio—a two-way communication technology that was independent of corporate or government control—in the 1950s and early-1960s cable was essentially unregulated. Many observers were therefore optimistic about the range of potential new channels beyond entertainment, from banking to voting to at-home instruction. High on the list of hoped-for uses was urban community development via minority "narrowcasting," social service delivery, and and two-way information exchange.[22] Yet just as a variety of individuals and organizations were beginning to see cable's potentials, commercial broadcasters noticed the threat it posed to their dominance (and to infant UHF broadcasting), particularly in large urban markets. During the Johnson administration, broadcasters lobbied the FCC to assert its jurisdiction over cable television.[23]

Pressure from broadcasters successfully catalyzed the FCC's freeze on the expansion of cable television between 1966 and 1972. In its decision, the FCC ruled that cable operators could not bring broadcast signals into any of the country's top one hundred markets, effectively slowing, if not halting, the entry of cable into urban areas. Cable is not, of course, regulated only at the federal level, but these regulations would significantly determine standards for the medium.

The freeze spawned calls from President Johnson, the FCC, and numerous other quarters for more research and policy analysis to investigate future possibilities for cable in urban areas. The period of the freeze, 1966 to 1972 (which, incidentally, were the years when preparations were under way to launch Landsat), is filled with rich written speculation about the future of the medium. The six-year theoretical discussion about cable as an instrument of social welfare brought together a collection of individuals and organizations that was, to say the least, unusual. From defense-oriented think tanks to public-access promoters, from MITRE to the Mafundi Institute, this diverse group would champion cable communications as an important tool for urban problem solving. Documentation of their enthusiasm ranges widely: there were official reports to the president and federal agencies; think tank studies commissioned by the Ford and Markle Foundations; articles for city managers in *American City;* underground video collectives writing for the alternative press; and writings from the ebullient FCC commissioner Nicholas Johnson.[24]

During the debates about cable, enthusiasm for the social potentials of "wired cities" and how best to harness these potentials were based to some extent on the technology as it was and to some extent on the technology as it might be.[25] A consequence of the FCC freeze on entry into major urban areas was that most analyses of cable as a community development medium were either speculative or based on limited uses of the technology in towns and neighborhoods such as Dale City, Virginia, and Waianae, Hawaii. Some analysts drew inspiration from the short history of continuing education and public affairs programming on public television (also known as ETV [educational TV] and ITV [instructional TV]). Still a new phenomenon, this form of broadcasting seemed to offer opportunities to create programs sensitive to minority concerns. The reports on cable's early promise that made reference to the early public television experience noted that while public television had limited channels, public-access cable would have no such contraints. Diverse populations would be able to learn from one another and, as a result, create "one happy electronic clan."[26]

As participants in policy discussions about cable came to agree that the medium would become a municipal (as opposed to a regional or national) technology, a few major regulatory questions dominated, both during the freeze and immediately after. In the context of setting standards, participants asked: Should cities, private companies, or nonprofit community organizations own the systems? Should cable be financed by individual subscription, by advertising, or by having program producers pay to transmit their messages? The most contentious regulatory issue related the legal definition of cable to other media: Would cable be a common carrier, like the telephone? A publishing medium, like the newspaper? A broadcast medium, like the television? Or was cable so unusual that some day it would be considered essential urban infrastructure and be regulated as a utility, like electricity and water?[27]

Looking to a new medium for salvation, rather than enforcing regulations on old technologies to better serve the public interest, is a pattern that has repeated itself many times in the history of American technology. Cable's relative newcomer status—the absence of an established cable industry, and its possible use for two-way communications—made it a source of hope. At stake in the cable debates was whether the new medium, like many of its predecessors, would evolve to become a network primarily for business communication, like the telegraph, a system primarily for entertainment broadcasting, like television, or a public urban infrastructure, like sewers and roads, with the potential to improve daily life for all. Like debates about the privatization of the internet three decades later, many participants felt passionately that early decisions about cable's regulatory framework would shape the medium's future capacity to serve the public good. Their rhetoric was optimistic but not deterministic; cable's potential to solve social problems hinged on society shaping its uses through communications policy.[28]

An overarching theme emerges. In the era of American urban crisis, as communications policy became aligned with urban social welfare policy in the minds of many participants in the cable debates, so, too—for a subset of those participants—it became aligned with national security concerns on the home front. Defense analysts and grassroots activists shared the goal of finding, through communication technology, new strategies for community development, racial integration, social service delivery, and "power to the people." But for many of the defense intellectuals, the social benefits of citizen participation, adult education, job placement, and even "power to the people" were understood within a framework of national security planning that was by no means incompatible with their training for warfare and their interest in control.[29] These experts' concerns about maintaining urban security were obscured by their rhetoric about promoting citizen participation, citizen-government communication, and urban development through cable. This was not new. Michael Latham has identified similar rhetorical strategies in use during the Kennedy administration, when a number of foreign policy measures were rooted in social science ideas about reducing counterinsurgency through modernization. In the early years of the war in Vietnam, for example, activities in the region undertaken as part of security initiatives such as the Strategic Hamlet Program, were concealed behind public claims about humanitarian development. Arguing that there is an essential continuity between development operations in wartime and peacetime, Latham views the Peace Corps program in similar terms. In a context of interdisciplinary war, a limited role for community participation served the larger goals of controlled modernization and civil defense.

A similar rhetoric was in operation in domestic urban programs: concerns about maintaining urban security were sometimes concealed behind a benevolent rhetoric about community participation and urban development. Indeed, substituting the words *America's inner cities* for *Vietnam* while reading Latham's account of modernization programs is striking. Samuel Yette, who performed a similar word substitution with documents from the era, concluded that several of Johnson's Great Society programs were "pacification programs" for the American ghetto.[30]

Linkages among communications, urban development, and national security on the home front are quietly present in the most influential and widely cited position papers on cable policy from this era. Three examples—from the Task Force on Communications Policy, the National Academy of Engineering, and the Sloan Commission on Cable Communications—reveal how the community of defense intellectuals and their understanding of the urban crisis as a national security crisis was integral to framing the earliest conversations linking cable television and urban problem solving for the national defense. Like studies of remote sensing and urban research, studies of communication research and inner-city development cannot be understood as an impartial set of scientific inquiries; they should be read as research grounded in the nation's security concerns. The ideas these analysts explored in these reports would influence the FCC's eventual ruling on cable that ended the freeze.

Cold War Social Policy: Three Reports on Cable

The Johnson administration placed great faith in the tools of social science to engineer a better urban America. During his time in office, Johnson assembled several multidisciplinary expert task forces and commissions to prepare reports on some of the nation's most pressing problems (there were ten in 1965 alone).[31] The best known was the National Advisory Commission on Civil Disorders, chaired by former Illinois governor Otto Kerner, with New York City Mayor John V. Lindsay second in command. Not as well known, but central to the cable story, was the Task Force on Communications Policy. In an August 1967 "Message on Communications Policy," the president began by saying that communications are a potentially powerful source for enhancing world understanding. Focusing on emerging regulatory issues based on technological changes, Johnson called for a comprehensive review of U.S. communications policy. He charged the task force with reviewing and recommending policies for a range of technologies that included domestic telephone service, international satellites, and cable television. The task force's report can thus be read as an important statement on the United States' strategic position in the communications world order.

Johnson appointed Eugene Rostow, undersecretary of state for political affairs (1966–69), an expert on U.S.-Soviet relations and the Middle East and brother of Walt Rostow, to chair the task force (the report is known as the Rostow report). Previously a professor at Yale Law School, Rostow specialized there in antitrust issues, but he also had written on urban renewal and the role of the law in city planning. (One of his observations on planning was that virtually all the most admired European and Asian cities were designed under authoritarian regimes.)[32] Other members of the task force included cabinet representatives from across federal departments and agencies, among them Robert Wood, undersecretary of HUD, James Webb, of NASA (represented by Willis Shapely), and Charles Zwick, director of the Bureau of the Budget. Leland Johnson resigned his post as director of the Communications Policy Program at the RAND Corporation (as required by RAND's conflict-of-interest policy) to serve as staff director for the research team. Leland Johnson had authored many reports on satellites, national security, and international development policy, and RAND became a center for cable television policy research during the freeze and after (supported by grants from the Markle and Ford Foundations and later the National Science Foundation). Much of the task of policy analysis for the task force was contracted out to staff at research organizations; RAND, MITRE, the Stanford Research Institute, and the Spindletop Research Center prepared background papers for the final report.

The focus for the task force was communication as a global phenomenon, which explains its extensive discussion of satellites and instructional media in developing countries. Yet if not originally charged with examining cable's potentials for the nation's internal security, the task force's final report linked government-sponsored communication research for international development to possible programs for American inner cities. The task force's recommendations about telecommunications delivering information to Communist and developing countries through satellite broadcasts and the Voice of America cannot be separated from its recommendations about wiring America's ghettos for news and information, even if the connection was not explicit. One dissenting voice argued that "we have earlier expressed the view that the chapters on domestic common carriers and television broadcasting are inappropriate for the report and not requested by the President's Message," but this member of the task force was outvoted by colleagues. Thus, alongside its recommendations for international satellite policy and the role of U.S. communications in Third World development, in a segment on the future of television the report discussed cable as a means to maintain the nation's internal peace, particularly for economic development in poverty-stricken areas such as South Central Los Angeles.[33] A vision of municipal communications channels to carry both citizen-produced programming and two-way services for job training and continuing education was a centerpiece of the task force's analysis.

Given the task force's charge to focus on U.S. communications systems in an international context, it is not surprising that both the task force's director and its research staff director had extensive experience with security matters, both national and international. The appearance of men with similar training (and in some cases the same men) to guide cable policy analysis with a decidedly domestic focus at the National Academy of Engineering and the Sloan Commission confirms the connections that existed between defense priorities and development in an international context and urban social welfare policy and cable television policy in a domestic context. A bevy of analysts brought expertise gained in studies of national security and international development to bear on cable policy on the domestic stage.

At the National Academy of Engineering (NAE), the Committee on Telecommunications prepared a report on communications technology for urban improvement—a report sponsored by HUD and the Departments of Commerce and Justice. Chaired by Peter Goldmark, president of CBS (former chief of staff for New York City's Mayor Lindsay and second to Frederick O'Reilly Hayes in that city's Bureau of the Budget), the committee included representatives of several federal agencies and organizations such as the International City Management Association and the National League of Cities, academics, senior management at major military contractors (e.g., General Electric and RCA), and scientists and social scientists from research institutions such as NASA's Jet Propulsion Laboratory, the Atlanta Urban Observatory, and the Urban Institute. Consultants included Konrad "Kas" Kalba, an independent communications planner, Robert Peters, of the Stanford Research Institute, and Charles Zraket of the MITRE Corporation. This large group sorted itself into topical study groups to examine a variety of issues ranging from traffic management to distance education to crime control.

The NAE report, *Telecommunications in Urban Development* (1971), provides insight into linkages between cybernetic views of cities as communication systems and the argument that communication technologies can help to alleviate urban problems. Citing Lewis Mumford and Jay Forrester, the authors framed urban issues in terms of systems, defining cities as information processing systems in order to necessitate a systems approach. Beginning from the "problem orientation" offered by systems thinking, the authors acknowledged that there would be no easy technological fix to complex urban social problems, and they thus recommended telecommunications as a single piece of a more comprehensive strategy for urban improvement that "might well be followed by, and combined with, those of technologists in transportation, sanitation, energy, utilities, construction, and other fields, in the study of the city as a highly complex organization to which the principles of modern systems engineering can be applied."[34] Like their colleagues on the president's task force, NAE's committee members proposed numerous pilot projects, including cable-based surveillance for community crime prevention and training programs for minority media professionals. These and related innovations based on communications technologies would not solve America's problems alone, but they could be a significant part of a solution.

With similar aims, the Alfred P. Sloan Foundation—which at that time was also funding research on urban information systems at several university campuses—in 1970 established its Commission on Cable Communications, appropriating $500,000 for a study of cable in the nation's urban areas. Headed by Edward Mason, dean emeritus of Harvard's Graduate School of Public Administration, the commission included Eugene Rostow; James Killian, former MIT president, science adviser to President Eisenhower, and past chair of the Carnegie Commission on Educational Television; Jerome Wiesner, special assistant for science and technology under President Kennedy and later president of MIT; Henry Rowen, president of RAND; William Gorham, head of the Urban Institute; Franklin Thomas, president of the Bedford Stuyvesant Restoration Corporation; Kermit Gordon, president of the Brookings Institution; Carl Kaysen, head of the Institute for Advanced Study and one of Eisenhower's former science advisers; James Q. Wilson, Harvard political scientist and former director of the Joint Center for Urban Studies; and several former mayors. Background papers were prepared by researchers such as Ithiel de Sola Pool, Kas Kalba, and the deputy director of the research staff, Monroe Price, a professor of law at UCLA and consultant for RAND projects on cable and the Task Force on Communications Policy.

The Sloan Commission's final report, *On the Cable: The Television of Abundance* (1971), framed the medium's potential for programming diversity in direct contrast to the limited spectrum of broadcast television. An introductory focus on the promise of communications technologies for alleviating the problems of America's inner cities set the stage for the document that followed. Like many reports about cable from foundations and think tanks in the early-1970s, the authors acknowledged that, "at this moment," the medium was "not remarkably impressive."[35] It was cable's potential, not its reality, that was most exciting. Like their colleagues on the presidential Task Force on Communications Policy and at the National Academy of Engineering, the commissioners proposed pilot projects in medical communications and continuing education to evaluate the medium as a tool for urban development.

The Sloan Commissioners concluded that creating cable infrastructure in urban communities should be a short-term national priority. But citing the work of colleagues at the NAE, they suggested that communications might in the long term slow the "cancerous growth" of cities and make possible the more decentralized future society that Norbert Wiener had envisioned years earlier, wherein many interactions could take place not face-to-face but over the cable.[36] Skeptical about potential costs if the infrastructure were to begin its urban life solely as a public technology, the commission recommended an evolutionary policy strategy. Initially, market conditions would allow cable to spread most rapidly. Then, the report suggested, once private companies had laid cable infrastructure across the nation, making it universally affordable, government could step in to remake cable into a common carrier; only at this later phase would cable become a municipal information network, delivering services such as job information and medical care.

While their ultimate recommendations differed in the details, each of the three study groups relied on the expertise of scientists and social scientists to craft a vision for the future of urban cable. In this future, "100 channels" was often touted, but cable was never discussed as the money-making entertainment system it eventually became. Rather, each report imagined an urban infrastructure developed to provide a variety of services to users in their homes or in neighborhood telecommunications centers, an infrastructure with capacity for two-way communication as well as instruction. Ending the isolation of ghetto dwellers, this mass communication system would provide a means of modernizing the quality of life for millions of Americans. The frequency with which these reports were cited by authors and organizations across the political spectrum, including city task forces contemplating a cable franchise, confirms the defense intellectuals' influence in setting the terms of the cable debates.

Ithiel de Sola Pool on Communications and National Security

The history of cable is not the story of a military technology transferred to the civilian world. Like the history of radio and the U.S. Navy as described by Susan Douglas, cable technology was developed and used by civilians for two decades before it sparked the interest of the defense establishment.[37] As cable made its way into cities, seemingly poised to become a mass urban phenomenon, defense intellectuals took on a central role in setting the agenda for cable policy and programming. A cybernetic image of cities as communication systems helped to shape their perception of urban problems as communication problems in city subsystems, with cable technology a potential solution. Beyond this initial framing, it is less obvious, at first glance, how these researchers' prior work on defense-related topics, and with methods of military origin, colored their view of the medium.

One way to understand these influences is to interpret this moment in cable history as an example of the transferability of analytic tools that Henry Rowen speculated about in his remarks to the Military Operations Research Symposium: cable offered social scientists an opportunity to apply to the domestic scene the findings from their ongoing research into communications, civil defense, development, and urban insurgency in an international context. Christopher Simpson has argued that studies of media and modernization in the period 1945 to 1960 cannot be understood as objective social science; rather, they need to be contextualized historically as research seeking to eradicate Communism and bolster international stability. This argument bears extending to the late-1960s and early-1970s. It becomes highly significant that top policy advisers on cable's role as a medium for citizen participation, citizen-government communication, and violence prevention included researchers such as Leland Johnson, Ithiel de Sola Pool, Edward Hearle, W. Bowman Cutter, and Herbert Dordick. All had completed studies of communications and development in countries such as Vietnam, China, and Colombia. Many of their studies were sponsored (sometimes classified) research, analyses that military and intelligence operations (and some Agency for International Development projects) would use as the basis for later interventions.[38] Thus, in the cable story we find leaders in international development work and propaganda research turning to studies of cable communications in the urban ghettos of the United States. This made sense in an institutional context in which their colleagues were drawing direct comparisons between insurgencies overseas and urban problems at home.[39]

A close look at the work of Ithiel de Sola Pool helps to trace how analyses of cable engaged with an ongoing debate of the era in international development policy and in U.S. urban policy: the debate about managerial versus participatory approaches to reform. Pool, who had studied political science at the University of Chicago, went on to work at the Hoover Institute for the Study of War, Revolution and Peace at Stanford, where he served as assistant director of the Program in Revolution and the Development of International Relations. In 1953, Pool moved to MIT, where he was associated with, and later headed, CENIS. Along with CENIS colleagues Guy Pauker and Suzanne Keller, Pool shifted some research attention from overseas development to domestic urban affairs.[40] While in Cambridge, Pool joined the Harvard-MIT Faculty Seminar on Arms Control (with Henry Kissinger, Lincoln Palmer Bloomfield, Walt Rostow, and Thomas Schelling, working on war gaming). In 1965 he helped to found MIT's Department of Political Science, where one of his colleagues was Robert Wood, the future HUD undersecretary. Pool also founded the Simulmatics Corporation around this time, which pursued both civilian communication research and also defense-sponsored projects, including contributions to the pacification program in Vietnam. Pool's applied social science work to 1970 encompassed an astonishingly diverse set of topics: propaganda and mind control, social networks, communications under Communism and Nazism, peacekeeping and deterrence, computer modeling of politics, Voice of America radio, and development studies.

In the early-1970s, Pool became interested in research on cable communications and domestic urban issues. He consulted for the Sloan Commission and the MITRE Corporation. He served with other defense intellectuals on President Nixon's Science Advisory Council panel on Science and Urban Problems and the Department of Defense's Defense Science Board.[41] He also pursued urban cable research on his own. Pool's writings on cable and urban affairs bear a striking resemblance to his research on democracy, media, and international development. Both addressed the central question: "Is there a conflict between the goals of democracy and stability?"

For example, a few years before undertaking his cable work, in a 1967 paper "The Public and the Polity," Pool articulated what he viewed as one of the fundamental questions of modern democracy: How can a mass society satisfy individual citizens' desires while maintaining a stable society? He described his firsthand experience of the situation overseas and its lessons for the American case. Whether in the Congo, Vietnam, or the Dominican Republic, social order depended on "compelling newly mobilized strata to return to a measure of passivity and defeatism from which they have recently been aroused by the process of modernization."[42] So, too, the movement for civil rights often came into conflict with American society's need to maintain social order; U.S. leaders, he observed, faced a classic dilemma. Pool's essay went on to consider the relationship between political participation and political stability. He suggested that, while sometimes the two were mutually exclusive, given certain conditions, they could support one another; the trick was to find this balance.

The balance that became a popular theory behind American urban administrative reform in the late-1960s and early-1970s was decentralization—the concept of dividing large urban areas into neighborhoods and making government more responsive to citizens at the neighborhood level. This could include creating outreach programs in neighborhoods, shifting decision making and control of resources to neighborhoods, and developing neighborhood institutions to replace the traditional citywide institutions. Decentralization was applied widely to improve urban governance, in particular the delivery of municipal services and the sense among residents of control over their lives. According to Robert Yin and Douglas Yates, decentralization was among the motivating principles of the Community Action Program and the Model Cities Program. Administrative decentralization differed from community control insofar as it brought government closer to the people but did not hand over fundamental power.[43]

Proponents of cable, including Pool, saw cable as a perfect complement to administrative decentralization. By understanding social services as information exchanges and citizen participation as an essential component of maintaining feedback in the city system, decentralization expanded the cybernetic view of cities. This idea gained currency in the federal government, where appointees such as Lita Colligan, assistant to the director in the Office of R&D at the U.S. Department of Health, Education, and Welfare, came to praise the possibility that cable would reduce the costs and increase the efficiency of information and referral services. Cable could not eliminate poverty or solve hunger, she acknowledged, but waiting in line at welfare agencies might become a thing of the past. At RAND, too, several researchers working on cable, including Robert Yin and William Lucas, would turn to study the decentralization phenomenon as they continued their communications research. They noted the vagueness of the term *decentralization*—that it could rhetorically serve to justify any number of actions; nevertheless they suggested, combined with other reforms, it might help to alleviate some citizen alienation.[44]

Understanding Pool's intellectual trajectory helps to make sense of how his later vision of cable communications as empowering citizens—specifically, his enthusiasm for ghetto cablecasting and decentralization—logically extended his earlier theories into new territory. The tension between political participation and political stability was a central point of conflict during the civil-rights era and the urban crisis, when new groups clamored—sometimes violently—for power, and government sought to offer it in a controlled way. In a later paper on cable and electronic democracy, "Citizen Feedback in Political Philosophy," Pool asked, "How does one weigh the trade-off between the public welfare that arises from being treated with respect and equality and that which arises from delivered outputs?"[45] Taking up the issues that Seymour Schwartz had also raised about the psychic benefits of participatory decision making versus the risks to decision outcomes, Pool observed that two-way cable communications might aid in making urban leadership listen to more of its citizens. Characterizing CATV as a neighborhood, grassroots medium, he enthusiastically suggested that citizen feedback sessions would reduce alienation and create opportunities for "black capitalism" (capitalism was an important stage of development theory). He did not, however, support electronic voting, arguing that uneducated voters would likely take the process too casually.

This frame for an appropriate structure to organize increased community participation while maintaining some hierarchical control characterized many defense intellectuals' speculations about cable, as well as other social welfare initiatives such as the Community Action Program. Pool was not the only one calling for active—yet simultaneously circumscribed—public participation. Like many of the defense intellectuals trying to reconcile top-down managerial and participatory approaches to urban reform, Pool proposed that politicians "morally committed to liberal values" be the ones to lead the way.[46] Paraphrasing Max Weber's views on mass political leadership, he argued for what he believed was an appropriate level of citizen participation: citizens could be brought into some decisions some of the time, but not every decision all of the time. That was what elected political leaders and their trained policy advisers (like himself) were for. According to this view, some forms of managerial control were necessary to get anything done. Daniel Patrick Moynihan, President Nixon's assistant for urban affairs, agreed with this view. Citing Jay Forrester in his acknowledgement that expert reformers often pointed to counterintuitive policy solutions understood only by a few elite social scientists, Moynihan observed that the experts might "lose a few friends" along the way, but future historians would likely forgive them.[47]

Pool's biographers have taken sides. Christopher Simpson, representing the more negative interpretation, has emphasized Pool's long history of military- and intelligence-sponsored work on deterrence, information flows, propaganda, simulation and gaming, and overseas media as part of a cold war, imperialist quasi-conspiracy (Simpson does acknowledge, however, that many of the communication researchers of the period viewed themselves as leftists). According to this view, Pool was an example of the breed of elite social scientists who rose to prominence on the heels of applied psychological warfare studies, concealed as basic research on public opinion and mass communications. This is a familiar vision of social science, social welfare, development programs, and the media as forms of social control (Charles Silberman's phrase for the phenomenon in the social welfare arena was "welfare colonialism"). The most extreme critics suggested that the ineffectiveness of social programs rooted in social science knowledge was deliberate, ultimately serving the needs and preserving the power of elites.[48]

Other interpreters of Pool offer a positive appraisal, emphasizing his interest in "technologies of freedom," (*Technologies of Freedom* was the title of his 1983 book). Lloyd Etheredge has characterized Pool as a social reformer, well-intentioned in his studies of the roles of mass media in international modernization and development, tough and realistic in his appraisals of reform from inside the system. According to this more charitable interpretation, Pool was an intellectual consumed by fundamental questions about how democracy works, and whose research program was dedicated to understanding how media might contribute to the spread of democracy in the United States and overseas.

These portraits of Pool may seem to be mutually exclusive, but in fact they are not. They qualify Pool for membership in the club of talented cold war intellectuals who believed that academic research should serve the national interest. While it is true that Pool pursued contract research for military and intelligence operations, as well as NASA, as a defense intellectual of the period with an interest in social welfare, much of his work was taken on with the best of intentions for spreading the benefits of democracy around the world, if in a somewhat paternalistic way.[49] Many of the civilian systems analysts and social scientists working for government did not consider themselves to be "militaristic" thinkers; many identified strongly with the groups that they sought to help, considering themselves social reformers searching for the most efficacious way to help people by working within the system.[50] As Edward Erath, president of the Los Angeles Technical Services Corporation, told the U.S. Conference of Mayors in 1966, many DoD and NASA scientists and engineers had a social conscience.[51]

Pool's allegiance to "the system" over ceding control to minority groups, stressing decentralization instead of community participation, is an approach found in many urban programs of the period. As Robert Yin has observed, community participation has a long history in America, and this history is rife with ongoing tension between efforts for citizen participation from outside as well as inside the system. Federal urban programs such as the Community Renewal Program, Community Action Program, and Model Cities were efforts to bring low-income and minority Americans into community participation, but only under terms set by local governments.[52] Speculations about cable would be framed in a similar way, calling for increased "community participation"— but only of a certain kind.

Wolves in Sheep's Clothing: Alternative Voices

Think tanks and advisory panels of defense intellectuals were not the only voices shaping public discussions about the fate of urban cable. In contrast to the alliances among elites that dominated efforts to transfer systems analysis, computer simulations, space age management, and satellite reconnaissance to city governments, the breadth of interest in shaping cable as a medium for community participation was striking. A range of "alternative voices"— united by their prioritizing social justice and citizen participation over market forces and calling for citizen activism to achieve the goal of positioning cable as an instrument of social welfare—were among the most outspoken on how the technology's creative potential held great promise for minority voices on the urban scene. The ACLU, the United Church of Christ, reporters writing in the *Nation,* authors of books such as *Guerrilla Television* (1971) and *Cybernetics of the Sacred* (1973), and public-access media centers such as Open Channel, were among the alternative voices mobilized during the cable freeze. All employed the rhetoric of a "window of opportunity" and expressed the conviction that two-way communications, citizen-produced programming, and social services delivered via the wired city held great promise for mitigating social inequality.[53]

These alternative voices shared three core beliefs with the defense intellectuals. First, they followed directly from the Kerner Commission in stressing that getting more minority voices into the communications industries would simultaneously encourage group solidarity and facilitate communications across social groups. Second, they focused on the absence of both an established cable industry and a fixed regulatory framework to suggest that there was a window of opportunity to create a new power dynamic. Making technology seem relevant, and urging minorities to become producers of programming, as well as consumers, took on new importance. The choice was posed in stark terms: minorities could continue to be passive recipients of still more entertainment channels from white America or they could seize an opportunity to shape a medium with potential reach beyond the modest audiences of the black press. Third, the belief that information equalled power, and the recognition of an information gap, motivated calls to increase access to the cable medium for minority citizens.

Together with the defense intellectuals, media activists promoted the concept of neighborhood telecommunications centers to serve as training centers and places to centralize the production of, and even the viewing of, community programming. Creating community cable centers in minority residential areas would respond to concerns that the majority of early cable subscribers were likely to be affluent. And it capitalized on the historical fact that, in low-income areas, the multiservice center as neighborhood institution dated back many decades. These diverse proponents of diverse programming feared the potentially monopolistic powers of cable operators. They argued that this public-interest potential could be guaranteed only through regulatory decisions to make cable a common carrier; in other words, the medium's potential as a platform for free speech was too great to let market forces decide cable's ultimate form.

Certainly there were some differences in perspective between the defense intellectuals and those who self-identified as media activists. When it came to advocating for citizen participation, for example, the former argued for a decentralized approach with strong managerial oversight; the latter preferred a less hierarchical interpretation of "citizen participation," although the precise structure for this participation was hard to pin down. Yet while some activists explicitly called attention to these differences of opinion—for example, Michael Shamberg described his Raindance video collective (publisher of *Radical Software* and author of *Guerrilla Television*) as a counterculture alternative to RAND (indeed, its name is a play on RAND)—these institutions and their perspectives on cable, while motivated by very different concerns, were less easily differentiated than one might suspect.

In fact, a close inspection of the many alternative voices on cable reveals heavy influence from, and infiltration by, defense intellectuals. A look at some of the most vocal alternative voices finds, for example, several men who contributed to policy analyses for the National Academy of Engineering, the Sloan Commission, the Task Force on Communication Policy, and RAND. Ben Bagdikian, of the Kerner Commission and RAND's Communications Policy Program, published *The Information Machines* (1971). Ralph Lee Smith, a staff assistant on the Kerner Commission, wrote "The Wired Nation," which appeared first as an article in the *Nation* (1970) and then, two years later, came out as a book. Monroe Price, a professor of law at UCLA who served as deputy director of the Sloan Commission's research staff and as a consultant to the Task Force on Communications Policy and to RAND, crafted *Cable Television: A Guide for Citizen Action* (1972) for the United Church of Christ. Peter Goldmark, chair of the National Academy of Engineering cable research program, went on to direct an experiment through HUD's New Rural Society program, an effort to use telecommunications technologies for economic development in rural areas and, by extension, for urban problem solving.[54] In other words, the boundary between defense research institutions and liberal activism was permeable.

The consonance of beliefs between the defense intellectuals and some media activists was further emphasized by the circulation of several individuals between institutions and also by collaborative work. For example, when the MITRE Corporation sought to make the "wired city" concept a reality, it hired Ralph Lee Smith to its staff and then convened two advisory panels of experts, among them Pool, from MIT; Lloyd Morrissett, from the Markle Foundation; Theadora Sklover, from Open Channel; Bernard Gifford, from RAND; Alvin Schorr, from NYU's School of Social Work; and W. Bowman Cutter, from the Urban Institute's Cable Television Information Center.[55] Similarly, many alternative voices moved from centers of media activism to consult for defense research institutions. For example, Richard Kletter, of the Portola Institute's Media Access Center, consulted for RAND on the NSF-sponsored project *Cable Television: Making Public Access Effective* (1973), recommending background readings such as *Radical Software* and later jointly authoring *Cable Television: Developing Community Services* (1974) with two RAND analysts. There was frequent movement among this cast of characters, all toward the goal of ensuring cable's capacity to improve communication among citizens and between citizens and government.[56] Publications such as *Urban Telecommunications Forum* included representatives of this diverse group on its editorial advisory board.

Conferences on urban and community cable offered still more opportunities for defense intellectuals, aerospace engineers, and media activists to come together in collegial conversation. For example, there was a conference at the Urban Institute, cosponsored by Black Efforts for Soul in Television and the Urban Communications Group (June 1971); there was a seminar on Urban Cable Television at the USC Center for Urban Affairs, cosponsored by the Center for Urban Affairs and the Mafundi Institute (January 1972); an "Urban Cable Symposium," at MITRE, was organized in conjunction with the Urban Institute Cable Television Information Center (October 1972); and smaller panels met at conferences such as of the 1973 American Society of Planning Officials.

At the joint MITRE–Urban Institute conference in October 1972, representatives of nearly every institution involved in the cable discussions were present. Among the participants were Amitai Etzioni (Columbia University), Lloyd Morrissett (Markle Foundation), Herman Kahn (Hudson Institute, formerly RAND), Ralph Lee Smith (MITRE), W. Bowman Cutter and Charles Tate (Urban Institute Cable Television Information Center), Joe Wholey (Arlington, Virginia, County Board of Supervisors and formerly of the Urban Institute), Ted Ledbetter (Urban Communications Group), Paul Vischer (Hughes Aircraft), Henry Geller (FCC), Thea Sklover (Open Channel), Jim Taylor (Watts Communications Bureau); Red Burns (Alternate Media Center), Bill Wright (Black Efforts for Soul in Television), Martin Evers (Dayton–Miami Valley Consortium), Kas Kalba (Minnesota Experimental City), William Knox (National Technical Information Service), Al Siegel (HUD Research Program), Kujaatele Kweli (telecommunications consultant for the National Urban League), Seymour Mandelbaum (University of Pennsylvania), Glenn Ralston *(Urban Telecommunications Forum)*, Robert S. Powers (special assistant for urban telecommunications, U.S. Department of Commerce Office of Telecommunications), W. D. Richards (United Church of Christ), Snowden Williams (Office of New Communities Development, HUD), Harold Barnett (Washington University), Herbert Dordick (NYC Office of Telecommunications, formerly RAND), and William Gorham (Urban Institute). Several senators, mayors, city planners and other local government officials, executives from the cable industry, and representatives from civic organizations (e.g., the National League of Cities) also attended. The panelists' presentations and discussions revealed much more by way of shared belief than disagreement. Their reasons for adopting a particular position might have differed, but by and large on cable policy questions these strange bedfellows were of one voice: if an effort were made to get minority voices in at the ground level, the cable medium might quickly become a potent force for empowerment.

Why Such Strange Bedfellows?

Why would defense intellectuals and media activists share such interests? Why would Arnold Klein from Public Access Information Resources in New York City tell an audience that he and colleagues found reports from MITRE and RAND useful in their work?[57] Why would RAND researchers praise *Radical Software,* the publication of an institution that called itself the counterculture RAND, and push for a community cable station in Watts on the model of Pacifica Radio, the most antiwar broadcaster in the nation? (This same ideal of a live cable network modeled on the Pacifica Radio network was also the vision of Raindance founder Michael Shamberg.) It is only in the context of theories of development and expanded definitions of national security strategy that these proposals make more sense. Recognizing the counterculture's view of the alternative uses of media for "liberation," and the defense intellectuals' self-perceptions (some might say false consciousness) as working toward that same goal, it becomes possible to understand why a vision of cable as a tool for improving citizen participation and social service delivery, particularly in inner cities, served the needs of both communities.

Chapters 2 through 5 described the construction of public problems and how conceptions of cities as cybernetic and large-scale systems satisfied the needs of multiple constituencies, from defense and aerospace contractors seeking new markets to city leaders seeking a scientific attack on urban problems. In the case of cable, a vision of the medium as a tool for citizen participation and municipal service delivery served an even more diverse set of interests. On the one hand, it fit well with the 1960s counterculture that viewed technology as an instrument for liberation. Enthusiasm in the counterculture for high technology so tightly linked to military and business interests was based on assumptions about how it might be used to support progressive causes. A concern for cybernetics' humane possibilities and a vision of cable as a tool for what political leftists defined as "good" as well as "evil," undergirded this view. For this reason, we find the *Whole Earth Catalog* praising Norbert Wiener's cybernetics, and an early electronic bulletin board in San Francisco calling itself Loving Grace Cybernetics without a trace of irony. Similarly, media activists such as Michael Shamberg and Paul Ryan called for the mobilization of "guerilla television" and "cybernetic guerilla warfare," reasoning that citizen-produced media were weapons to fight the "perceptual imperialism of broadcast television."[58]

Simultaneously, a vision of cable as a tool for increasing community development and racial integration also served the defense intellectuals, who perceived themselves to be liberal reformers, not agents of military domination. Thus, in 1968, RAND researchers Ben Bagdikian and Kathleen Archibald proposed that Los Angeles create community programming and that it include a "televised ombudsman." With this format, residents of the inner city would be able to voice concerns, and local groups such as the Sons of Watts and the Venice Gangbusters would be able to follow up on their complaints. Such a self-perception even characterized Herman Kahn, a RAND analyst who went on to found the Hudson Institute, architect of Mutually Assured Destruction and one of the inspirations for Stanley Kubrick's Dr. Strangelove. When the MITRE Corporation hosted a conference on Urban Cable Systems, it was Kahn who gave the keynote speech, telling the audience about his upbringing in a gang-infested ghetto and his empathy with inner-city Americans. "I, Herman Kahn, white, Jewish if you will, educated, has more empathy with the Negro parent in the ghetto than you middle class Negroes with black skin."[59]

Despite the defense orientation of much poverty research, then, many social scientists of this era tended to be politically liberal or leftist, especially when young, and had a sense of empathy with the people they sought to help through social programs. This empathy also characterized New York City's Mayor Lindsay, who so often sided with the city's poor and racial minorities that his reputation among the middle classes was damaged.[60] According to this view, the alliances that emerged between defense intellectuals and media activists during the six-year FCC freeze on creating urban cable systems were not so out-of-the-ordinary in this period.

Occasionally, however, this "false consciousness" came to light. While interactions in fora devoted specifically to cable generally were collegial, interactions between defense intellectuals and community activists in other fora were less so. For example, during the 1969–70 academic year, antiwar protesters overwhelmed participants at one of a series of symposia on computers, communications, and the public interest that was hosted jointly by Johns Hopkins University and the Brookings Institution. Participants included Paul Baran, Martin Greenberger, Leland Johnson, Nicholas Johnson, James Coleman, Eugene Fubini, Alan Westin, Ralph Nader, Kingman Brewster, Daniel Bell, Karl Deutsch, Lee DuBridge, Richard Posner, Herbert Simon, Larry Roberts and Clay Whitehead. The session featuring Anthony Oettinger, Alain Enthoven, and Ithiel de Sola Pool had to be rescheduled on account of the confrontation.[61]

Similarly, when RAND researcher Robert Yin—who did not come from a defense background—visited Yale to discuss the possibility of co-teaching a course with Douglas Yates in Yale's Urban Studies Program, his visit was not warmly received. The proposed course was to focus on New York City neighborhoods and urban policy, and Yin "always imagined our participant-observation work as being street level research," attuned to citizen perspectives. He was therefore shocked to find that outsiders perceived the New York City RAND Institute as an agent of the Lindsay administration, an enthusiast of methods limited to quantitative systems analysis, and an embodiment of the "Establishment," unreceptive to real community needs.[62] The course was not offered.

Not all of Yin's colleagues were surprised by criticism, however, as can be seen from a comment on an internal memo written by Roberta Goldstein. Goldstein gave her boss an account of the fallout that occurred after she shared with leaders at the Brookdale Community Mental Health Center some of the details of RAND's research for the Community Mental Health Board. One administrator's reaction to her talk was particularly hostile: in his view, RAND "was the civilian arm of the military, involved in counterinsurgency," and he reported that black activists in the community thus wanted nothing to do with it. Attached to the memo in the RAND Corporation archives is a cover note from New York City RAND Institute head Peter Szanton (July 22, 1969): "I am surprised that a year and a half went by before the first such incident occurred."[63] The note reveals the tension that pervaded many of the era's efforts at urban reform. Activists charged that a war was being waged on black America, and yet elites widely believed that military management methods would blaze the trail toward equality in urban areas.

A turning point for the history of cable television came in 1972. The cable freeze began to thaw that February, when the FCC, in its report and order, laid the foundations for the next generation of cable policies. Many reports and much lobbying from the defense intellectuals and their activist colleagues achieved some success in embedding citizen participation and citizen-government communications into policies shaping the emerging medium, at least in theory. New regulations required cable operators in large metropolitan areas—the top one hundred markets—to set aside three public-interest channels for public access, municipal government, and local program origination.

Conventional accounts of this moment in cable history, both by critics of the period and in more recent analyses, often portray it (like so many of the Great Society programs) as a story of failed efforts to create a new medium for citizen participation. According to this view, the FCC's mandates did not go far enough, either because cable did not become a common carrier or because the rulings did not stipulate how public access and municipal service programming would become realities. Yet when compared to radio and commercial television policy, the FCC's policy on cable represented a major victory. Even under private ownership, urban cable systems would be required to carry programs created by local citizens and local governments.

Understood in the context of cold war social science, the 1972 report and order becomes exactly the opposite—a major success for defense intellectuals in shaping social policy. Cable was not a military technology, yet visions of its uses to improve government decision making, municipal services, and the democratic process were tightly bound up with an expanding conception of national security strategy. It was in no small part due to pressure from defense research institutions that public access and municipal government channels became the standard in urban cable policy.

Extending the cybernetic view of cities as communication systems, the defense intellectuals' frame for the future of big-city cable fundamentally shaped American urban history—even before cable systems were commonly franchised at the municipal level. Cable became conceptualized as a local medium for urban communications, in contrast to commercial broadcast television in the network era. Wide-ranging conversations about cable policy popularized an understanding of cable networks as fundamental urban subsystems. Cities such as Los Angeles soon made explicit the analogy to streets, sewers, gas, and other infrastructure networks, handing control of cable networks to their Bureaus of Public Utilities. Now all that remained was to transform rhetoric and regulations into reality. Chapter 7 investigates efforts to implement recommendations about cable in the years following the report and order.

CHAPTER SEVEN

Wired Cities

In his 1972 remarks at the MITRE Corporation Conference on Urban Cable, just months after the report and order, Herman Kahn offered his thoughts on the future of wired cities: "If I had a guess, I would say this kind of TV will not be successful in removing the alienation or in education or in changing the minority groups."[1] Kahn's views were decidedly out of step with the views of his think tank colleagues. The few efforts that were under way during the freeze to develop municipal cable systems as virtual community centers, voting systems, and universities—most notably in New York City—seemed to indicate a promising future ahead. When cable franchising for large cities finally became a realistic possibility in the early-1970s, and public officials began to make critical decisions about opportunities for citizen-produced programming, Kahn's colleagues enthusiastically took on new roles as cable advisers to city governments.

"The biggest single difficulty I would say of the kinds of study which MITRE, Hudson, RAND do," explained Kahn, "is that they are done by people who are not business-oriented, who don't understand about quick marriages."[2] Defense intellectuals had played critical roles making public-access and government channels part of cable policy, yet Kahn forecast they would encounter significant roadblocks implementing their ideas. During the six-year waiting period, American cities experienced many changes—a new presidential administration, reductions in federal funding, and continuing middle-class migration to the suburbs, with consequent loss of city tax revenues.[3] In this context, Kahn suspected, local government priorities vis-à-vis cable might have shifted.

In fact, Richard Nixon's 1969 entry into the White House had signaled a sea change in many aspects of urban affairs. The administration's first official act was to establish a Council for Urban Affairs in the cabinet to formulate new approaches to urban problem solving. In the face of continuing mismatches between what social science claimed it could do and what social policy actually accomplished, the large-scale federal funding for social programs such as the War on Poverty began to dry up. Johnson's Great Society initiatives largely were dismantled. As Daniel Patrick Moynihan, assistant to the president for urban affairs, remarked when discussing leadership in urban affairs in the post-Johnson era of lost faith in government spending for problem solving, there was no "Admiral Mahans or George Kennans to provide a master theory of an urban policy comparable to previous formulations of foreign and defense policy."[4] Instead, the Nixon administration placed a new emphasis on private institutions and public-private partnerships.

As Kahn predicted, these tectonic shifts at the federal level, and their aftershocks in urban coffers, would have implications for cable franchising. Combined with a nationwide recession in the early-1970s, the political and economic climate of urban areas in the period following the report and order differed significantly from what it had been only a few years before. Without large-scale funding for infrastructure development on the order of a military operation, city leaders became far less interested in the civic potentials of the new medium and far more concerned about capturing the financial benefits of a franchise.

Had the cable freeze not lasted for six years, had cable infrastructure development become a large-scale government operation of the Johnson or Nixon administrations, the history of cable as an urban technology might have come closer to the defense intellectuals' visions of wired cities. In the market context of the early-1970s, however, wired cities could not survive as originally conceived. Mayors such as New York City's Lindsay had praised the long-range perspectives that think tanks offered as an "anti-bureaucratic tonic." Yet the experiences of several cities illustrate how, as the 1970s went on, reports prepared during the cable freeze, with their emphasis on community participation and social services via television, came to seem increasingly out of touch with the needs of both urban administrators and the urban poor.[5] American cities were eventually wired, but cable's early potential as a forum for political debate, adult education, and job information was eclipsed when new regulations set a course for the renegade technology that remade it into an arm of existing broadcast empires.

New York City

The move to create community cable alongside efforts to bring top-down management innovations to New York City paralleled Mayor Lindsay's theories about improving urban administration. Upon taking office, Lindsay first concentrated on increasing centralized coordination and reducing duplication through a reorganization of city government. Soon after, his office began to implement some "necessary duplication" toward the goal of decentralized community planning. Piggybacking on the Community District Planning Boards established in the 1963 revision to the New York City Charter, the city's decentralization program included Community Planning Boards, Little City Halls (also called Neighborhood City Halls), and Urban Action Task Forces. Anecdotal evidence suggests that these formal programs and Lindsay's personal style combined to give New York City's minority citizens a feeling of increased participation in city operations.[6] Cable communications were intended to have a complementary effect.

Like the residents of mountainous rural areas, where cable originally was developed to solve a technical problem, for years New Yorkers watching television had suffered poor signal reception due to their city's many tall buildings. This occurred despite New York City's being home to the three major broadcast television networks. To address the problem, in 1965 the city granted temporary franchises to three cable operators, becoming the first major urban area to embrace the medium. The city's decision was ideally timed; when the FCC froze cable expansion in large cities the following year, New York City was grandfathered. Its cable companies were not required to suspend operations.

Thus, as researchers for the FCC and for President Johnson were preparing reports about the future of urban cable during the near-nationwide freeze, they had the great fortune to look to the experience of New York City—the nation's most populous urban area—for guidance. Understanding how cable unfolded there helps to explain why national hopes could run so high for the medium to become an instrument for urban community development. In New York City, public-access and municipal information programming were centerpieces of the city's vision for cable from early in the planning process. This vision was first made public with the release of a city task-force report on the future of cable in September 1968.

Following Lyndon Johnson's lead, Mayor Lindsay created his own advisory task force—the Task Force on CATV and Telecommunications—in 1967. Its assignment was to study the potential form and uses for cable as it related to other city communication systems. Fred Friendly, former president of CBS News, then professor of journalism at Columbia University and television adviser to the Ford Foundation, served as chair. Friendly had resigned from CBS in protest over the low level of programming. He believed that by not offering enough serious journalistic coverage of the world, commercial television had reneged on its commitment to the public interest. His concern with the public interest is reflected in the task force's report as well as in Friendly's later personal criticisms of the implementation of cable. The task force report (September 1968) argued that cable, if regulated in an appropriate way, would offer "the most promising solution to a number of difficult communication problems facing our larger cities."[7] Lindsay was urged to quickly develop a citywide system. The report also proposed making public access a requirement of all New York City franchise agreements.

The task-force recommendations were rapidly implemented as city policy. At first, the city extended its temporary contracts with the original three cable operators, and those operators continued to wire the city and to experiment with ways to improve reception. On July 29, 1970, in a more permanent decision, the city's Bureau of Franchises granted twenty-year franchises to two of the cable operators: Sterling Manhattan (Sterling, which eventually became Manhattan Cable), owned by Time-Life; and TelePrompTer Manhattan (Teleprompter), of which Hughes Aircraft, a prominent player in the early days of cable, owned 49 percent. (The third operator, CATV Enterprises, had had technical difficulties; it was not granted a franchise in 1970.) These contracts offered direct evidence that city leaders viewed cable as a neighborhood medium. The city required each operator to divide its territory into ten smaller districts to facilitate local community programming. It stipulated that each city neighborhood had to be wired equitably. And the city's director of franchises, Morris Tarshis, setting a precedent that would shape the future for cable, required that there be channels for public access.

Two years before the FCC's report and order, New York City's contracts required that each cable operator provide two public-access channels, one that could be reserved and one that could not. To further encourage access by a multiplicity of voices, the city imposed limits on screen time: individuals and groups could air for no more than seven hours per week, no more than two of these hours in prime time. Cable operators were instructed to set aside two channels on July 1, 1971, and another two in August 1973. By 1973, it was expected, each cable operator would be offering a minimum of twenty-four channels, with several of them specifically for public and city use.[8]

And so, during the years that most other large cities were banned from cable franchising, cable infrastructure development in New York City moved forward. Four channels for citizen participation became available from July 1, 1971. With access to portable video equipment (Sony had released its CV video portapak in 1968), theoretically any individual or group could have a say on the city's cable systems. Despite the excitement, however, early analyses found that public-access television for most urban residents remained more an ideal than a reality. Little use was made of the public-access channels, and surveys revealed that most nonprofit organizations were not even aware of their availability. This was in sharp contrast to Canada, with its tradition of government-funded media: in major cities there, citizen cable got off the ground quickly.[9]

New York City (unlike Canada) did not take government action to promote the use of the new cable infrastructure; rather, several local private foundations took a leadership role, seeding the start-up of several new organizations to facilitate citizen cable programming. Most prominent among them was the Markle Foundation, which in cooperation with New York University (NYU) created an Alternate Media Center on the campus in April 1971. George Stoney (an American trained in Canada) and Red Burns (a Canadian) headed the center and its ten salaried employees. A few months later (July 1971) the Markle Foundation and the local Stern Fund together supported Theadora Sklover in establishing Open Channel, another "facilitator organization" that could offer technical assistance to groups wishing to make use of the cable channels. With its twelve salaried employees, Open Channel worked with schoolchildren as well as adults. These and several other organizations became critical intermediaries for getting citizens' voices on the cable.[10]

Thus, as the FCC was preparing its report and order, a massive experiment was under way to test ideas about how local, citizen-produced programming might alter the quality of urban life. David Othmer's *The Wired Island* (1973), a report commissioned by the Fund for the City of New York, offered a balanced appraisal of the first two years of public-access cable television in New York City. Othmer acknowledged the slow start to public access, the reality of much dull and poor quality programming, and continuing inequities in access to the medium. Yet the medium's imperfections did not appear to deter individuals and groups from seeking access to production equipment.

Othmer's account charted a steady increase in programming from 1971, when citizen channels first became available. By June 1972, more than 650 original hours (60% of them from ten major user groups) had appeared on both systems. Comparing the two cable operators, Othmer observed that Teleprompter had done far more to encourage citizen programming than Sterling. For example, in 1972 Teleprompter opened a storefront public-access viewing studio on 125th Street in Harlem, which also lent production equipment. Original programming on Teleprompter increased from under fifty hours per month in the first year to about two hundred by the end of 1972. By mid-1973 the figure was well over three hundred hours per month. Sterling, by contrast, in its first year restricted access to its facilities, producing most of its "local programming" outside its studios. However, during the early summer of 1972, Sterling agreed to help establish a venue for public access in cooperation with the Alternate Media Center of NYU and the City of New York. This Video Access Center opened in the West Village near NYU. Maxi Cohen served as director, aided by an all-volunteer staff. With production equipment available for borrowing and a video studio and viewing room, the center opened on September 15, 1972, to a standing-room-only crowd. Patronage continued to be brisk.

The cable medium, particularly on the Teleprompter system, seemed to be realizing many of the goals that diverse groups had laid out for citizen participation and community development. There were three main types of users: individuals, facilitators such as the Alternate Media Center and Open Channel, and organizations established for other purposes using the medium to communicate their message. Among them, such city treasures as Mount Sinai Hospital, in the Teleprompter service area, began to make use of cable to deliver social services (in the Sinai case, for health information) from the fall of 1973.[11]

According to analysts of the day, the most robust aspect of the medium was its function as a community communication device. Local organizations that created programming in cable's early years were diverse. They included, among others, Vietnam Veterans in the War, the Society for the Prevention of Drug Addiction, the National Organization for Women, the Inwood Committee for Irish America Action, the Film Maker's Cooperative, Friends of Haiti, the Center for the Analysis of Public Issues (which focused on programming for deaf New Yorkers), and block associations. On one occasion, when residents of Greenwich Village were debating the location for a new school, the meeting was transmitted via cable so that people could participate at a distance.[12] Both analysts and users began to claim that public-access television in New York City offered a new public space analogous to city streets. In Othmer's account, watching public-access programming closely resembled "visiting Times Square. It is exhilarating, frustrating, shocking and boring—above all, it is simply amazing."[13] The uses for cable seemed to be closing in on realizing the fantasy of a "wired city." Citizens with diverse perspectives could interact via the cable medium, and hopes rose that protest and disagreement would take place not violently in city streets, but peacefully, mediated by cable communications.

Yet while the early years of cable in New York City were successful in generating interest among citizens to produce programming, serious questions about audience size remained. Among the impediments to attracting committed viewers were not only the technical limitations in program quality and reception but also problems of building penetration, in part due to landlords. Othmer, distinguishing cable from the commercial broadcasting that many media activists identified as being a form of psychological warfare, proclaimed that "Public Access, by its very nature, cannot be systems-analyzed or 'preprogrammed.' Its aim is to give the medium back to the people."[14] His remarks raised a question that would be asked frequently: Did the medium serve the public interest merely by allowing citizens to express their views, or did there have to be a regular audience before cable enthusiasts could declare success? On one of the channels it was not possible to develop fixed time slots, and this made for difficulty in developing a repeat audience and, by extension, regular outlets for extended conversations.

In the midst of the New York City experiments, the FCC released its report and order on cable policy. Some proponents of cable as a community development tool were disappointed by the decision not to subject the medium to federal common-carriage requirements and not to make it a municipal technology, like a utility, everywhere. Yet, unlike radio or broadcast television, cable was the only medium of its day that was required to carry multiple channels for citizen access and local government use.

Shortly after the report and order, an activist facilitator organization, Survival Arts Media, proposed a citywide celebration in New York City in April 1972 to honor the first anniversary of public-access cable programming. A three-day Public Access Celebration was held in July, the goal being to celebrate the success of citizen engagement with public channels on the medium and to increase viewership. In conjunction with the event, Sterling and Teleprompter were interconnected, and Teleprompter opened its Harlem studio. Approximately twenty temporary viewing centers were also created. In keeping with the event's activist roots, the celebration also offered opportunities for political action on cable policy. Franchise requirements for equitable wiring lacked enforcement, which meant that many community groups that had created programming were unable to view it. To showcase the continuing lack of access to citizen programming in certain neighborhoods, for example, during the celebration, one news program was interrupted by an announcement that residents on the Lower East Side had "nothing to celebrate." For the next twenty minutes, in solidarity the screen went blank.[15] Thus, at the same time that New Yorkers were feting public access, many celebrants were pushing for more.

Following the festivities, organizers of the celebration issued a report to document the event and to present their recommendations for the next phase of cable development and regulation. They called for public hearings on the use of public channels; they proposed that use be free; they urged that more access facilities be created; they argued that more experiments in innovative programming should begin; they suggested that viewing centers be created in schools, museums, and other public facilities; they insisted that requirements for cable operators to hook up poor areas be enforced; and they instructed the city that a committee of citizens should advise its Office of Telecommunications.[16]

Defense Intellectuals in New York City

Among the recommendations for further reading made by the organizers of the Public Access Celebration were documents ranging from *Radical Soft-*

ware to RAND Corporation reports. But how, beyond appearing in bibliographies, did the defense intellectuals participate in the New York cable story? Not, interestingly, under the auspices of the New York City RAND Institute, although its vice president, Edward Blum, did author several papers on municipal information systems.[17] One prominent defense intellectual, Herbert Dordick, found a home at the city's new Office of Telecommunications, part of the Bureau of Franchises. Dordick, formerly of RAND, was hired to head the city agency, one of many new government offices created by administrative reorganizations under Mayor Lindsay. As a RAND consultant, Dordick had authored reports on systems analysis and logistics, on education planning in developing countries, on telecommunications policy research, on advanced sensing techniques, and on other space research for NASA. He had also worked on engineering in regional development and had served as lead author on *Telecommunications in Urban Development* (1969), one of the staff papers for President Johnson's Task Force on Communications Policy.

The Office of Telecommunications was the first of its kind in the nation (the decision to create it was made in 1971, although it did not open until July 1972, the one-year anniversary of public-access cable). In the creation of this new bureau we see, as in Los Angeles' CAB, the city's recognition of the importance of a new class of technologies as fundamental to city infrastructure. The office had four official functions: "Enforcing existing contracts with the two cable operators; franchising the remaining boroughs; developing uses for the two city channels; and formulating overall policy for cable."[18] The bureau had access to city lawyers and three city planners from the planning department, and Dordick predicted that, as cable penetration and programming grew, the staff would expand to include engineers, technicians, planners, and programmers. Yet even this recognition, and Dordick's long-standing interest in telecommunications for urban development, were not easily translated into the broad range of imagined uses for the medium. Although the Office of Telecommunications publicized the availability of channels for citizen-produced programming and even helped to make cable programming accessible for viewing in the city's housing projects, it was unable to adapt cable to improve communications between New Yorkers and their local government. Even though the franchise board's requirements had created several channels for use by the city, and even though the city's experience with government television went back to 1961 (when the FCC created WNYC Channel 31 as an experiment to train civil-service employees), the municipal channels were never programmed.[19] The much-vaunted potential for citizen feedback to government was never realized. Dordick coordinated the office for only a year, leaving for a job as professor of communications at USC, in Los Angeles, where he could pursue his continuing interest in cable infrastructure for citizen-government communications. In his letter of resignation, Dordick however claimed several victories for the office, including having increased the use of the public channels fivefold, made cable accessible to all housing projects, highlighted the role of telecommunications planning as an important urban infrastructure, and promoted New York City as the pioneering city for cable communications.[20]

A second place where the influence of defense intellectuals on New York City's cable infrastructure was evident was at the John and Mary Markle Foundation. Markle seed money for facilitator organizations such as Open Channel was largely responsible for making citizen participation possible. Although research funding does not itself create ideology (money cannot do research), in making choices about whose research to fund, what research to fund, and in some cases in seeking out researchers, foundations and federal sponsors do shape the trajectory of knowledge.[21] Without the substantial assistance of the Markle Foundation, whose director, Lloyd Morrisett, was a prominent defense intellectual, citizen cable programming in New York City likely would not have been possible.

In its first four decades, the Markle Foundation exclusively had funded medical research. With the appointment of Morrisset in 1969 (he served through 1998), the foundation shifted its interest to communications research, with a focus on communications technology and "the public interest." Morrissett moved to Markle from the Carnegie Corporation of New York and its allied Carnegie Foundation for the Advancement of Teaching, where he was vice president (James Killian was president). During the cold war, Carnegie, like Ford, was an important sponsor for communication research, area studies, and other social science with ties to military and intelligence operations. Morrissett began to take Markle in a complementary direction. Under his leadership, Markle seeded many cable facilitator groups and also research on cable communications at MITRE, the Urban Institute, and RAND. Morrissett was well connected into the networks of leading cold war social scientists, serving as a trustee at RAND (he was chair of the New York City RAND Institute Board of Trustees), at the System Development Corporation (and its affiliate, the System Development Foundation), and at the Children's Television Workshop, which he helped to found. He was also a member of the Council on Foreign Relations.

Markle's funding for cable research at MITRE, the Urban Institute, and RAND, as well as Dordick's later work in Los Angeles, persisted in trying to fashion the medium into a form that would complement larger efforts to secure the nation's domestic peace. The Markle Foundation's commitment to developing urban cable infrastructure in this image, expressed through its funding for urban cable research and programming in New York and in other cities, and its role actively contracting specific studies (in addition to simply funding those who applied for grants), continued for several years.

How much of an impact did these defense intellectuals have in New York City? Each of the above examples illustrates a case of indirect as opposed to direct influences on policy and programs. In their commissions and task forces and reports on cable's future, defense intellectuals had laid the six-year theoretical groundwork to make cable infrastructure into a municipal information system. First, they imagined a future incarnation of cable and then they helped to align federal communications policy with that vision. Yet unlike in military organizations, where one can expect orders from the top to be swiftly implemented using all available resources, city governments did not approach these experts' recommendations, or even federal communications policy, with that kind of commitment.

Cable's early uses in New York City appeared to demonstrate the medium's potential as an instrument for community communications, if not for citizen-government communication, increasing hopes for creative cable use elsewhere. A press release from the National League of Cities/U.S. Conference of Mayors testified to New York City having set the standards that the FCC later adopted; and it said that further collaborations with NLC/USCM would continue the city's pioneering efforts to bring cable into urban problem solving.[22] Yet around the time that the FCC issued its guidelines for urban cable, New York City's financial fortunes were shifting. Mayor Lindsay still referred to the medium as an "urban oilwell," yet other accounts suggested that the short-term economic viability of cable television in large cities was far from assured. In New York City, early successes attracting citizens to cable were not matched by financial gains for the city. That same year, Dordick and NLC/USCM's Frank Young told an audience that "there is no light at the end of that tunnel for New York" and that "cable may be the SST of telecommunications."[23]

In 1974, when Abraham Beame was sworn in as New York City mayor, the city was on the verge of bankruptcy. In this economic climate, the earlier goal of fully wiring the city dropped from the public's list of priorities. Limited funds for social services were better spent directly on professional service providers than used for experimentation with cable programming. By 1976, a report from the Urban Institute's Cable Television Information Center described how the most innovative social programming in New York City (as elsewhere) had been discontinued.[24] A single large cable operator, Warner Cable, bought out both TelePrompTer Manhattan and Sterling Manhattan. Warner, an arm of the Warner entertainment empire, would continue to invest in developing the city's cable infrastructure so long as it could turn the medium into a revenue-generating entertainment enterprise. Like many other city dwellers in this era, New Yorkers found both their city government and their cable operators turning away from the earlier sense that cable would become a medium for public-access and social service programming. A variation on this story played out in Los Angeles.

Los Angeles

Los Angeles, home to several think tanks and a large aerospace industry, was the nation's epicenter of film production and its second center for television. Yet Los Angeles was far slower than New York City to embrace cable television as part of its urban infrastructure. As early as December 1, 1965, the League of California Cities issued a model ordinance for cable communications, signaling the medium's entry into Los Angeles and other California cities. City records show that many small companies submitted applications to operate local franchises; in 1966 alone, Community Cablecasting petitioned to operate a system in Pacific Palisades; Harriscope Inc. sought to operate a system in the Universal City–Toluca Lake area, and Pacific Master Systems applied to operate in the City of LA. Yet in 1976, a decade later, cable reached only 6 percent of homes in the Los Angeles area.[25]

Part of the slow pace of cable development can be explained by the FCC freeze, and part of it can be explained by geography. Los Angeles' decentralized physical form, with many minicenters dispersed across a wide urbanized area, was ideal for civil defense planning in the 1950s. But for cable operators two decades later, it was prohibitively costly to create a metropolitan infrastructure. Instead, cable companies focused their efforts on a few smaller communities within greater Los Angeles, such as Santa Monica, home to RAND and SDC. The cheapest areas to wire would have been the dense ghetto areas, but cable operators were deterred, believing that those neighborhoods offered an unlikely subscriber base.

As in New York City, the idea that cable might become part of the city's future municipal infrastructure was first brought to the attention of city leaders by expert task forces. The task forces were not devoted specifically to cable communications; they were advisory panels on communications and adult education, part of a citywide Los Angeles Goals Program. This municipal initiative—which overlapped with the FCC freeze—was the pet project of Calvin Hamilton, the city planning director. While planning for the program was under way as early as 1964, Hamilton turned it into part of the city's response to the McCone Commission's *Report on Racial Violence in Watts* (1965), an investigation of the factors behind riots in the city in summer 1965. The McCone Commission was named for commission chair John McCone (who had no prior experience in urban affairs but who had distinguished himself in a variety of military and intelligence posts).

Hamilton created the goals program soon after his appointment as planning director in November 1964. Though itself a local effort, it deepened the city's commitment to many of the aims of the federally initiated Community Renewal Program (CRP) and related redevelopment initiatives. The goals program was among the most comprehensive outreach efforts by a city to develop citizen interest and representation in the city planning and management process. City officials heard from citizens in a variety of fora ranging from opinion surveys to the new Citizen Goals Council. From citizen recommendations and the recommendations of its expert advisers, the city eventually created a metropolitan master plan.[26]

The goals program, like many urban programs of the era, was managed from above, top down, while simultaneously incorporating citizen participation. When implementation began in April 1965, prior to inviting contributions from average citizens, the city assembled several expert issue-oriented "goals committees." These committees, comprised of specialists on issues such as poverty reduction and traffic flow, included a more general "council"—the Los Angeles Goals Council, made up of local leaders. Participation by nonexpert citizens came later, in September 1967. Among the promotional events to spark citizen interest was an information circus at the Los Angeles Civic Center, in June 1967, at which Hamilton pantomimed a speech with assistance from clowns holding cue cards. Drawing responses similar to those aimed at federal urban programs, the goals program ran into criticism from two directions: on the one hand, it was said that it wasted time and money consulting and engaging with average citizens who knew little about city management; on the other hand, it was said that it did not have enough citizen participation.[27]

It was during the early phase when experts were invited to provide technical and administrative leadership for the LA Goals Program that RAND's participation began. City leadership asked RAND researchers to consult on two specific aspects of the program: communications goals and adult education goals. RAND's Herbert Dordick, who later would direct New York City's Office of Telecommunications, authored reports on both topics for the City of Los Angeles in 1968, one with colleague Leonard Chesler. In the reports, he made cable a central focus.

Dordick noted that, before the freeze, several companies already had begun the process of obtaining cable franchises. Observing this interest, he advocated that the city step in and make a deliberate investment in wiring impoverished neighborhoods so that the medium could be used for community development. In Dordick's words, cable television could be "made the voice of the ghetto dweller" for communication within and among neighborhoods and between neighborhoods and city government.[28] Citing New York City's WNYC-TV to suggest that there were precedents for his ideas, he said that, nevertheless, Los Angeles could improve upon these earlier efforts. Dordick envisioned that the LA Goals Program, charged with physical infrastructure planning and committed to an understanding of poverty as in part a result of social isolation, could also assume responsibility for the large-scale planning of a municipal information network. He suggested a two-step process, beginning with the use of traditional television and moving to a cable network built and operated by the city. In his cybernetic view of Los Angeles and its problems, such efforts to improve city communications infrastructure would result in the reduction of information gaps, facilitate adult job training, and in turn improve race relations throughout the metropolis.

Dordick's Los Angeles reports are the origins of the enthusiasm for urban cable pilot projects that he expressed in his later staff paper for President Johnson's Task Force on Communications Policy. Cooperation between community groups in Los Angeles around a different medium, radio, by good fortune provided some of the evidence for this enthusiasm. For in Los Angeles, organizations such as the Mafundi Institute, a black cultural center in Watts, and Pacifica Radio's KPFK, an alternative radio station, already were collaborating to create local radio programming. Specifically, it was Pacifica's 1969 creation of a walk-in studio at the Mafundi Institute that inspired Dordick to envision analogous fora for two-way cable communications. With his RAND coauthors for the national task force, Dordick would propose pilot programs for cable communications in two low-income and minority neighborhoods, Watts and South Central Los Angeles.

This was a paradigmatic example of the strange coalitions that developed around cable as a medium for citizen participation. Analysts from a defense-oriented research institution actually praised the Pacifica Radio Foundation, which had previously been accused of being a Communist-controlled medium (in December 1962 and January 1963 the Senate Subcommittee on Internal Security held closed hearings on the matter).[29] It is only by taking into account the era's expanded definition of national security strategy that these proposals for the future of cable make sense. Reducing the alienation and improving the welfare of urban populations—goals of the Pacifica network—had become critical components of domestic security planning.

In New York City, where Dordick temporarily relocated in the early-1970s, city dwellers put communication principles into practice as soon as access to production equipment and studios were made available. Los Angelenos' experience with cable was noticeably different. For a variety of reasons—dispersed geography, lack of technical interference with broadcast television signals, existing community radio programming, limited funding for facilitator organizations, and a city council reluctant to grant cable franchises—the introduction of cable communications as a medium for local programming came far more slowly.

For example, the Mafundi Institute, home to the Watts Communications Bureau, was one of cable's early community-based enthusiasts in Los Angeles. Mafundi was created in 1966 with support from the Communications Foundation and the Kettering Foundation; Don Bushnell, president of Communications Associates, Inc., and a former consultant for SDC, was a founding member. The institute grew out of the Watts Happening Coffee House, which was created jointly by local Unitarian and Presbyterian churches following the rioting. In addition to its walk-in studio for KPFK, the bureau operated a center to train young people, including school dropouts, in basic video filmmaking. The bureau also worked with members of the Watts Writers Workshop to create video productions based on their writings. Lacking any easy way to display these programs to large audiences, staff at Mafundi became interested in setting up a neighborhood cable network. In a petition reflecting the ideal of cable as a community communication device, the Watts Communications Bureau asked the city for a cable franchise for South Central Los Angeles, one of the earliest facilitator organizations to do so. Despite repeated petitions and a long drawn out battle with the city, they never succeeded. In the late-1970s, the city officially rejected the bureau's franchise petition.[30]

As efforts to expand the Los Angeles Goals Program continued, RAND analysts pursued several other lines of research linking new media to citizen participation. A study of local political candidates' uses of cable television in advance of 1970 elections in Waianae, Hawaii (a Model Cities designee) concluded that there were many benefits from using the new technology in a small community.[31] The following year, Leonard Zacks and Craig Harris prepared a more theoretical report on a cable-based direct democratic legislative structure for local government. Attributing one cause of urban rioting to poor communications between citizens and government, Zacks and Harris went on to explore how cable might improve the representation of minority groups and their interests in local government. They proposed that cable serve as a medium to realize the community participation provisions of federal antipoverty programs. Acknowledging that prior attempts to help minorities gain access to political power had failed, their solution was direct democracy via cable. The estimated cost to build a municipal system for Los Angeles? $60 million.[32]

The state of affairs in Los Angeles was very much on the minds of a multitalented group that met at the Center for Urban Affairs at the University of Southern California (USC) in January 1972. Presenters at the conference included mayors, city councilors, RAND and Urban Institute researchers, representatives from the Aerospace Corporation and Hughes Aircraft, research staff from the Sloan Commission, planners from Model Cities agencies, senior administrators from the White House Office of Telecommunications Policy and the New York City franchise board, representatives from network television and the National Cable Television Association, academics, and activists from Mafundi, the Portola Institute, and Pacifica Radio. The meeting, cosponsored by USC's Center for Urban Affairs and the Mafundi Center of Watts, with additional sponsorship from the Brotherhood Crusade, Cypress Communications, the Stern Fund, and the Urban Coalition of Greater LA, convened just days before the FCC issued its report and order.

In his overview of the conference, researcher Robert Warren expressed an unusual optimism about cable's potential. Dismissing ongoing frustrations with other defense and aerospace tools applied to city decision making, Warren envisioned a bright future for American cities adopting the cable medium.[33] The conference proceedings were a standard exposition of the widely held conviction that cable should be a public utility, a community tool, and a new voice for minorities. Transcripts of the meeting and examination of the list of participants offer further evidence of how defense intellectuals and urban activists were in frequent and cooperative conversation (similar links were evident at an Urban Institute meeting several months before and at a MITRE meeting later that year). Ted Ledbetter, a black engineer and founder of the Urban Communications Group, an organization dedicated to advancing racial minorities' interests in the cable industry, publicly acknowledged several familiar faces in the room.[34] Morris Tarshis of New York City's franchise board echoed Ledbetter's sentiments about constant dialog among these strange bedfellows, noting that they seemed to be "riding the circuit in the things we talk about."[35]

These comments hint at the major accomplishment of the cable debates. This accomplishment was not, as promised, the creation of municipal cable systems. Nor was it, as predicted, an end to the isolation and alienation of inner-city residents: all of this talk did not translate into action, even after the 1972 requirements for public-access and municipal channels. Rather, the most significant outcome from the years of ongoing discussions about cable was the creation and maintenance of social networks where the defense intellectuals came together with big-city mayors and private foundations to sustain their high profile as urban reformers.

The creation of cable infrastructure in Los Angeles post–February 1972 proceeded at a snail's pace. Statistics show that in summer 1972, cable in Los Angeles was highly decentralized, with service provided by Theta Cable of California (whose parent company was Hughes Aircraft) to a subscriber base of 26,000. This was barely 1 percent of the city. In Beverly Hills, the same company, Theta, had only 3,980 subscribers. By contrast, cable penetration in Santa Monica (home to RAND and SDC) was far higher: Theta had 44,000 subscribers. Compton had no franchise, nor did Watts—even though the Watts

Communications Bureau had petitioned the city council to create one. Most frustrating to cable's advocates, the public-interest uses of the channels were negligible, and by 1976 little research on viewership had been undertaken. A report prepared that year for the transportation and public utilities department examined information sources and media usage in various Los Angeles minority communities and looked at the sorry state of public access. It found that public access in the city was limited to a single channel on the Theta Cable system. Theta operated a small studio in Santa Monica for citizens' free use, but its facilities were tapped to produce only four to eight hours of original programming per week. Further complicating the realization of a wired city was that the public-access channel reached only 70 percent of Theta's subscribers; it was available only to areas of the system with fourteen or more channels, and reception additionally required the use of a converter.[36]

A similar story emerged for government channels. In the few areas that were wired, government channels were not being used at all. In the case of local debates about zoning, for example, despite public complaints by planning officials that few residents attended neighborhood planning meetings, the city chose not accept the cable operator's proposal to remedy the situation by using a government channel.[37] Further impeding the realization of a wired city, the existing cable franchises, even though owned by the same company, did not interconnect, and this made citywide cable an impossibility. The much-touted uses of the medium for citizen feedback had little relation to reality.

In remarks to the 1972 USC conference, Thomas Bradley, president of the LA City Council and, from 1973, mayor, had expressed some urgency about cable franchising. There were "some time bomb aspects of the whole business of cable television," much like "the time bomb aspects of our urban communities," he had noted. Bradley suggested that the city act quickly to ensure that the public interest was protected.[38] The city officially confronted the issue of comprehensive planning for cable infrastructure at the metropolitan level in mid-1973, when a Citizens Committee on Community Antenna Television was formed. After further delay, in April 1975 the Board of Public Utilities and Transportation appointed an official citizen committee to advise the board and the city council on creating a master plan. Like the Los Angeles Goals Program, the aim was to combine expert and citizen opinion to create the blueprint for an LA master plan for cable.

The choice to identify cable as a public utility and as part of a master plan seemed to suggest that city leaders considered the medium a potential future metropolitan-wide infrastructure. So, too, did their choice of Dordick as coordinator of the LA Cable Communications Planning Task Force. Dordick, returned from New York to USC in 1973, was still promoting this view of cable in his academic and other contract work. (He was also still using occasional military analogies in his cost-benefit analyses of cable systems, comparing options for citizen-government communication by citing the "knapsack problem," or "submarine problem.")[39]

Los Angeles, however, like cities across the nation, began its official study of cable in an economic climate that would quickly turn the discussion away from the medium's potential as an urban public good. City officials reported to Dordick that only with outside financial support would they consider doing a pilot project focused on cable's citizen uses. City monies were limited, and they were needed for more urgent urban priorities. Nevertheless, Dordick persisted. His allegiance to cable as an instrument for the public interest led him to continue to advocate for municipal information infrastructure on an older model of universal service. His final report for the Bureau of Transportation and Public Utilities, submitted in 1976, presented a vision of citywide cable that echoed his reports to the Los Angeles Goals Program from 1968, proposing innovations such as a cable studio in City Hall and franchise districts mapped onto city neighborhoods. It also included, like recommendations from his former RAND colleagues Robert Yin and William Lucas, a call to use cable as a means to decentralize government agencies and services.

Despite his continued advocacy efforts, Dordick acknowledged that until more citizens recognized cable as an important public good and gave it a high priority, the city would be unlikely to lay out the $600 million (ten times the figure proposed by Zacks and Harris) that he estimated to be necessary to get a citywide network off the ground. His report was unable to reach an optimistic conclusion, acknowledging the equally low probability of a private system operator finding it financially viable to serve poor areas of the city. The "wired city" looked to be an increasingly unlikely proposition.

Beyond the Largest Markets: The Role of Think Tanks

The experiences of New York City and Los Angeles guiding the development of urban cable infrastructure were not entirely typical. As the nation's centers of media production and two of its largest broadcast markets, in both cities the variety of individuals and organizations engaged in efforts to make urban cable a medium for citizen participation was unusually large. Political leaders at the highest levels, from Lindsay, in New York City, to Bradley, in Los Angeles, praised the social possibilities of cable television alongside its potential financial returns. In another respect, however, efforts to create cable infrastructure in these urban areas were typical: both cities experienced a rapid shift around 1973, when across the nation enthusiasm for the medium as a public good began to dwindle. As urban administrators were turning toward a view of cable systems as a possible financial boon, a number of smaller cities were just beginning to look into franchising their own municipal systems. In this era of transition, city leaders were asking: What should these new systems look like? Should they embrace any aspects of the "wired city" so popular only years before? Researchers at several think tanks, who had been considering these questions for several years, stepped up to offer guidance. And in the efforts that followed, many close working relationships developed between city administrators and the defense research institutions whose staffs had played central roles in developing cable policy during the freeze.

At three think tanks—RAND, MITRE, and the Urban Institute—defense intellectuals, now prominent in cable research, persisted in their efforts to shape urban cable during the implementation phase. These institutions were not carbon copies of each other: RAND focused on policy questions, MITRE on technical standards, and the Urban Institute on hands-on franchise decision making.[40] When think tanks created guidelines for franchising and offered recommendations for programming, many cities (e.g., Cincinnati and Jacksonville) explicitly modeled their decision-making process, and in some cases their eventual cable systems, on these guidelines and recommendations. Reports from the think tanks became important sources of information for local decision making, as did publicity about their work in cities such as Dayton and Reston.[41]

The think tanks' continuing commitment to cable in many cases was made possible through support from the Markle and Ford Foundations, two private foundations whose sponsored research nurtured the idea that national security expertise should play a prominent role in addressing the nation's complex urban problems. Ties between the think tanks and these sponsors signal continuing efforts to apply findings from cold war social science research to cable development. Both Markle and Ford merit recognition for their sustained roles in funding early research on cable policy and later cable pilot projects directed by think tanks and media activists. That each sponsored both defense intellectuals and community activists to work with cable confirms that they—like many of the defense intellectuals—did not see the two constituencies as working at cross-purposes.

Ford had been a long-standing supporter of social science applied to national needs. It had created and sustained area studies centers across university campuses in the 1950s and urban affairs research centers in the 1960s. It had made RAND independent from Douglas Aircraft and it had helped to create both CENIS at MIT and the New York City RAND Institute. It had sponsored the Conference on Space, Science, and Urban Life and research investigating the transfer of military and aerospace technology to urban areas. A month before the report and order, Ford gave $2.5 million to city and state governments to help develop cable television. Ford's president, McGeorge Bundy, like his counterpart at Markle, Lloyd Morrisett, had prior involvements with educational television, instructional television, and public television. Also like Morrisett, as well as H. Rowan Gaither, Ford's former president, Bundy belonged to the social networks of defense intellectuals. A former Harvard professor and dean with ties to CENIS, Bundy was one of the architects of the war in Vietnam during his time on Lyndon Johnson's national security team (serving as special assistant for national security affairs and national security adviser). Bundy left the Johnson administration in 1965. He took the reins at Ford in 1966, heading the foundation through 1979. According to Leland Johnson, Bundy's interests significantly shaped the Ford Foundation's profile.[42]

These foundations and their presidents helped to perpetuate hopes for using cable as a cold war technology. Their support—intellectual, financial, and political—for cable experiments directed by RAND, MITRE, and the Urban Institute's Cable Television Information Center, strengthened links between defense research institutions and American cities. Yet as the 1970s went on, efforts by think tanks to shape the development of cable toward increased urban communication and community participation, and away from entertainment and profit, seemed increasingly out of step with the times.

RAND Comes to Dayton

Although in Los Angeles, RAND's successes implementing cable were limited due to the local geography and local politics, outside of that city RAND was recognized as a leader in setting standards for municipal cable policy. It was in a RAND-sponsored research study that Harold Barnett and Edward

Greenberg had coined the phrase *wired city* that quickly entered the main-stream.[43] Indeed, beginning in the late-1960s RAND became the intellectual center of cable policy research, thanks to funding from Ford, Markle, and the National Science Foundation. RAND staff rarely served as hands-on consul-tants—they preferred to stand back, study the issues, and make recommenda-tions.

Following the report and order, RAND's Communications Policy Program published several general reports, among them *Cable Television: A Handbook for Decisionmaking* (1973) and *Cable Television: Citizen Participation after the Franchise* (1973). In two cities, however, RAND staff chose to pursue a more di-rect advisory approach. Spartanburg, South Carolina, in a project under the direction of William Lucas, became a pilot site for experiments related to adult education and day care (the NSF program sponsoring the effort was Research Applied to National Needs—very much in line with how many defense intel-lectuals imagined themselves and the function of social research). And when city officials in Dayton, Ohio, began to consider creating a cable franchise, they called in RAND's communications team.

Dayton's experience hammering out the details of cable policy with assis-tance from RAND offers an especially interesting example of how expert social scientists' impressions of their own politics were not always mirrored by the people with whom they interacted. The background to the Dayton project was the cable freeze, during which the Miami Valley Council of Governments (comprised of fourteen cities and villages) declared a moratorium on franchis-ing. Funding from the Ford and Kettering Foundations gave Dayton the lux-ury of inviting seven RAND staff to examine issues ranging from cable infra-structure financing to regional social geography to system ownership alternatives. From the outset, city leaders expressed interest in how their fran-chising decisions might eventually contribute to the medium's potential to improve the quality of urban life.

Dayton had experienced urban riots in the 1960s and had a high percent-age of black residents in its central city. West Dayton had been designated as one of HUD's Model Cities, funded beginning in 1969. Other communities in the region varied widely in terms of economic resources and social needs. Thus, a central question for RAND's study became: Should a county-wide fran-chise be granted or would smaller franchises work better? RAND's researchers sought to design a cable system that could address area minorities' communi-cations needs—and that also was financially viable. Researchers were increasingly aware of the difficulties of creating sustained community programming (RAND's Ned Feldman had written about its early failures in Dale City, Virginia, and Lakewood, Ohio). Recognizing that the "public interest" must be locally generated, Robert Yin surveyed Daytonians about their information needs. The surveys found blacks to be more interested than whites in using cable communications to access job information and other social services. In January 1972, the RAND team invited municipal officials and community groups to offer further recommendations at a two-day "Policymakers Conference on Urban Communications."[44]

RAND's good intentions toward Dayton's minority residents did not match the reality of those citizens' expectations, however. The experts' recommendations about what would be the most viable cable system conflicted with what, to many Daytonians, seemed to make sense. The system that aligned best with what racial minorities said they needed was a dedicated, central-city franchise. According to RAND's quantitative analyses, such a system was not likely to be economically survivable, and in its final report the RAND team recommended that the franchise not be granted by the city of Dayton but by the entire region. In other words, the report proposed the creation of a metropolitan system.

Negative response came swiftly. James McGee, the city's Democratic mayor, saw RAND's plan as a recommendation that subscribers in the densely populated inner city in effect subsidize the more affluent and less densely populated suburbs. RAND staff suggested that the only way to guarantee service to Dayton's inner city would be as part of a metropolitan system, but many of the city's prominent minorities asserted that RAND's final proposal was an affront to their interests. With backing from McGee, the city rejected RAND's conclusions for a single-owner metropolitan system. The Dayton episode serves as one example of how RAND's perception of its own liberal sensitivities did not always match up with local realities. Subsequently, RAND's Communications Policy group chose not to work directly as franchising advisers to other city governments.[45]

Instead of following RAND's recommendations, a group of residents from the primarily black southwestern area of Dayton formed a joint-venture company with California-based Cypress Communications. The new company, Citizens Cable Corporation, proposed that Cypress build a network for the entire city of Dayton, including the black southwest area. In the southwest area of the city, Cypress would cede 50 percent ownership to local black citizens groups. This system became the third franchise in the country under black ownership. By 1975, a grant from the U.S. Department of Labor's Concentrated Employment Training Act of 1973 (CETA) to the Cable-Communications Resource Center in cooperation with the Miami-Valley Consortium and the Dayton Board of Education had created a program to train Dayton-area low-income residents for jobs in the cable industry. The program was administered by the Garfield Skills Center and after two years had placed a remarkable 92 percent of its trainees.[46]

MITRE Comes to Reston

RAND was not the only defense research institution to take on urban concerns. During the Johnson administration, the MITRE Corporation—like the Institute for Defense Analyses, System Development Corporation, and Hudson Institute—responding to increasing federal spending on domestic programs, created a civil systems division at its second office in MacLean, Virginia. MITRE, a nonprofit systems engineering firm, was established in 1958 in Bedford, Massachusetts, as a spinoff from MIT's Lincoln Laboratory. Like the RAND contingent at work on the SAGE air defense system that would be spun off to form SDC, a contingent from Lincoln Labs Division 6 went from work evaluating SAGE and general aerospace defense to work for MITRE. MITRE's leadership, a mix of men with ties to the Ford Foundation, RAND, SDC, the Institute for Defense Analyses, MIT, and the military, places the institution squarely within the networks of defense intellectuals discussed throughout this book. It included H. Rowan Gaither (chair of the board at RAND and the Ford Foundation), Julius Stratton (MIT president from 1959 to 1966 and a board member at both RAND and Ford), Franklin Collbohm (a RAND founder, former president, and board member at both RAND and SDC), William Webster (trustee of RAND and MIT and also former chair of the DoD's Research and Development Board), James McCormack (a vice president for research at MIT, a retired air force general, and founder of the Institute for Defense Analyses), and James Killian (president of MIT from 1948 to 1959 and President Eisenhower's science adviser).[47]

Like RAND, in its early years MITRE primarily contracted with military sponsors, first and foremost the U.S. Air Force. Also like RAND, during the conflict in Vietnam MITRE sent staff to Southeast Asia, opening an office in Bangkok for defense-related work. And again like RAND, MITRE's prominence in military communications work, developing the SAGE air defense system

(1958) and the National Military Command System (1963), as well as its shift to civil systems research, were likely reasons that MITRE staff were invited to prepare work for President Johnson's Task Force on Communications Policy.

Defense intellectuals' vision of cable as a "cold war technology" in most cases centered on uses for the medium in the context of the era's expanding definition of national security. They imagined that the medium would provide for the national defense indirectly, by reducing the citizen alienation that precipitated violence. This was certainly true of researchers at MITRE, but with one important addition. At the same time that MITRE staff in MacLean were exploring cable's potential as a delivery medium for urban social services, they were considering linking their civilian system, called TICCIT, into the DoD's ARPANET. Simultaneously, colleagues in Bedford were investigating the possible use of cable as a high-speed means for secure military communications. MITRIX, as this network system was known, might have become a technical standard, had researchers at Xerox Parc not created the much cheaper Ethernet at about the same time.[48]

Thus, while MITRIX never became a significant military communications system, MITRE's urban cable research must be situated in the context of the institution's simultaneous work to create a high-bandwidth defense communication network. At MITRE, as at RAND, then, ties were close between national security research and urban cable research. The difference was that while RAND staff focused on policy connections, MITRE staff were investigating physical connections.

At the MacLean office, MITRE's urban cable research revolved around TICCIT (Time-Shared, Interactive, Computer-Controlled Information Television), a two-way cable communication system. TICCIT (sometimes TICCET, with the *E* standing for Education) did not actually originate at MITRE. The system was imported in 1968 from C. Victor Bunderson, at the University of Texas. Bunderson, an education researcher, originally envisioned it as a computer-assisted learning tool, a computer-controlled, two-way instructional television system.

Directed by Kenneth Stetten, head of computer systems at MITRE, researchers initially sought and received funding from NSF to further develop an educational system. But when funding priorities at both the Markle Foundation and the NSF turned to emphasize how telecommunications might serve as a medium to deliver social services, MITRE shifted TICCIT's focus from education to social service delivery. Arguing that "urban planners, social scientists, and historians are increasingly concerned that one-way TV communications may be adding to, rather than ameliorating, basic problems of our society," MITRE researchers proposed to develop two-way cable communications as the centerpiece for a future urban communication infrastructure. With this infrastructure, they speculated, "new forms and imaginative uses of telecommunications" could "make contributions of fundamental importance to meeting and solving nearly every major problem of urban society and life," including ameliorating alienation and binding together communities through communication.[49] This rhetoric successfully courted both Markle and NSF, and thus research began on two-way cable communication projects for citizen-government communications and communication among government departments. Each of these investigations remained theoretical until an agreement with Reston, Virginia, opened opportunities to test them in a small-town setting.

Reston, a planned community with a population of about ten thousand located close to MITRE's MacLean office, was a privately developed analog to the kinds of small towns promoted in HUD's New Communities Program in the 1970s. Like the Community Renewal and Model Cities programs, which similarly employed the language of scientific experiment, proponents envisioned that the New Communities Program (and their privately developed counterparts) offered unparalleled opportunities to remake local government institutions, including using cable infrastructure to improve the political system.[50] The political situation of unincorporated Reston was complex. In its early days, it had no real local government beyond a homeowner's association; other decisions largely were in the hands of a private developer. A small cable station already was operating, and plans were under way to partner with the University of Virginia and Virginia Polytechnic Institute to expand opportunities for instructional programming.

MITRE staff took a leadership role, and in July 1971 Reston's demonstration cable system became operational, connecting a few homes and schools in Reston to MITRE's computer in MacLean through the Reston cable system, or RTC (the Reston Transmission Company, which had been involved since May 1971). Since the CATV system in Reston was not originally equipped for two-way communication, a portion of the system was retrofitted by the Jerrold Corporation. TeleVision Communications, Inc. (part of Warner Communications, and RTC's parent) paid for this. The small-town context made it possible to concentrate on experiments with cable communications specifically to and from individual homes, as opposed to neighborhood telecommunications centers.

As research and development went on, demonstrations of the TICCIT system were offered for public viewing. The setting for this demonstration was a home, where a MITRE representative showed how to operate the system. (Researchers remarked that a different set of demonstrations, with children in a school setting, were far more realistic in their interactions.) Between 1971 and 1972, visitors came from think tanks, government, and the private sector; reports characterized the typical visitor as male, forty-five, white, and professional. Demonstrations were well enough attended by representatives from President Nixon's cabinet that a 1973 report from MITRE concluded "we have reason to believe that the demonstration also influenced the Office of Telecommunications Policy and the FCC toward adapting a policy encouraging two-way cable development."[51] The evidence does not confirm this, nor is it weighed in relation to the New York City experience; nonetheless, that so many high-level visitors (listed by organization in the back of one MITRE report) attended suggests that the MITRE group's work was widely perceived as significant. Imagining the future of the Reston system from May 1973, the researchers projected that subscribers would soon be able to call up the central database through the telephone to request specific services.[52]

Early on, in April 1971, as the TICCIT system in Reston was under development, the Markle Foundation approached MITRE with a proposal for the staff's next project. Assuming the Reston experiment would be a great success, the foundation invited the research team to lay the groundwork for a far larger experiment: a study of nearby Washington, D.C. (population seven hundred thousand). The centerpiece of the analysis, as requested by the Markle Foundation, was "to identify the demographic, social and economic needs of urban areas which can be addressed by improved telecommunications services."[53] MITRE's preliminary report on the project, in 1971, *Urban Cable Systems*, included numerous economic and demographic analyses of Washington D.C., alongside discussion of the technology's social potentials. Based on this work, MITRE became one of several recipients of an NSF grant the following month, and further funding from Markle, to develop a more sophisticated analysis of cable's potential to deliver social services in the capital.

For assistance in thinking about how to make the "wired city" concept a reality, MITRE hired Ralph Lee Smith, author of *The Wired Nation*, to its staff. The research team convened two advisory panels of experts, including Ithiel de Sola Pool, Lloyd Morrissett, Theadora Sklover, Charles Tate, and W. Bowman Cutter.[54] In October 1972, the team cohosted with the Urban Institute a conference on urban cable systems, bringing together the wide range of cable enthusiasts for a demonstration of TICCIT and several days of discussion.

MITRE's final report in 1972 (like the earlier publication, titled *Urban Cable Systems*) predicted cost and demand for services and outlined system requirements for infrastructure development in Washington, D.C. Staff compared both the affordability and social utility of several alternative system designs. Based on the ongoing experiment in nearby Reston, authors concluded that cable offered possibilities for improving urban life in the capital. They noted that an evaluation study of the frustrations New York City encountered in getting cable off the ground would need to be completed to ensure success in developing a "socially useful" system for Washington, "an illustrative example of a city with serious urban problems and a declining quality-of-life."[55] The plan that eventually emerged was for a Washington Cable System, divided into multiple districts supported by nine cablecasting studios across the city, with service areas following Mayor Walter Washington's recent administrative divisions of the capital. MITRE proposed making this system a reality in time for the bicentennial in 1976 to be a showpiece for the nation.

Yet by 1974, evidence from the Reston experiment did not augur well for the project in Washington. In Reston, as in larger cities across the nation, small, independent cable operators like RTC were on shaky financial ground. The limited financial backing for cable experiments from sources such as NSF and Markle was directed to research institutions such as MITRE to sponsor analysis, not to private companies like RTC to pay for infrastructure development. That year, Warner Cable took control of the Reston cable system.[56]

Nevertheless, it appears that MITRE's influence may have helped to sustain Warner's early commitment to citizen-produced programming, even if the two-way capacities envisioned by MITRE staff for municipal social services did not become operational. In a special issue of the anthropology journal *Human Organization,* devoted to media anthropology, anthropologists Martin Topper and Leigh Wilson described their work helping to set up a community news and feature program on the Reston cable system in order to mobilize citizen participation. According to observers, the local cable news show became an invaluable source of community information, particularly in circulating political information that Reston's private development corporation did not want to be made public.[57]

The community news efforts were short-lived, however. And MITRE's Washington, D.C., project never moved on to the next phase of development. By the mid-1970s, the cable system operating in Reston no longer was considered particularly spectacular, at least according to newspaper commentaries. Its chief improvement on broadcast television was not advanced interactive services but some local programming (twenty-four-hour weather, a movie channel, a tickertape news channel).[58] By 1976, a study prepared for the National Science Foundation reported that MITRE's Reston's experiment had been suspended.[59] That year, MITRE gave TICCIT to the Hazeltine Corporation, which later sold it to Loral and Ford Aerospace. MITRE's vision of a wired city had become financially impractical, and the corporation retreated from civilian cable research.

The Urban Institute

The Urban Institute has never been a defense research institution in the narrow sense; from the start, its focus was urban problem solving. Yet in Paul Dickson's 1971 analysis of the history of American think tanks, the Urban Institute is juxtaposed alongside RAND, SDC, and the Institute for Defense Analyses. The institute's roots and its early leadership make the case. Established in 1968 at the request of President Johnson and explicitly modeled on RAND, its first proposed title, the Institute for Urban Development (which signaled ties to development theory) was eventually rejected in favor of the Urban Institute because the acronymn for the first-choice name would have been IUD.[60] The Urban Institute's first president was William Gorham, who had worked at RAND for nearly a decade. Gorham was one of McNamara's "whiz kids," working at DoD (deputy assistant secretary of defense), and then the Department of Health, Education, and Welfare (assistant secretary for planning and evaluation), where he brought PPBS—planning, programming, budgeting systems—into use in the late-1960s.[61]

The institute's early board reflects the era's conviction that expertise in defense and international security matters could translate to domestic social welfare matters. In addition to Robert McNamara (secretary of defense under Kennedy and Johnson), trustees included Eugene Fubini (assistant secretary of defense and deputy director for defense research and engineering in the Pentagon under Johnson), Cyrus Vance (deputy secretary of defense under Johnson and secretary of state under Carter), William Ruckelshaus (later acting director of the FBI and head of the Environmental Protection Agency under Reagan),

Joseph Califano (who had worked in the Office of the Secretary of Defense from 1961 to 1964 as assistant to McNamara before becoming Johnson's special assistant for domestic affairs), Edward Levi (president of the University of Chicago and, later, attorney general under Ford), and William Scranton (special assistant to the secretary of state under Eisenhower and vice chair of the National Advisory Panel on Insurance in Riot-Affected Areas under Johnson).

Like RAND and MITRE, in its early years the Urban Institute was largely federally funded, although rather than lean on DoD it found primary sponsors in HUD, OEO, HEW, and the DoL. The institute also received support from the Ford Foundation. Like RAND and MITRE, the Urban Institute engaged with a spectrum of individuals and institutions concerned with the quality of urban life, from operations researchers to community activists. It did so in a variety of ways: through an urban fellows program, which brought civil-rights leaders into year-long working relationships with the staff of analysts at the Urban Analysis Group; by distributing reports from activist organizations such as Ted Ledbetter's Urban Communications Group; by encouraging staff such as Harry Hatry, director of the institute's State and Local Government Research Program, to serve on the editorial boards of journals such as *Operations Research;* through cosponsored events—with MITRE on urban cable television in 1972 and with the Washington Operations Research Council on urban growth and development in 1973; and simply by dint of sharing a building, 2100 M Street, with RAND's Washington, D.C., office.[62]

In June 1971, the Urban Institute hosted its first conference on cable television, cosponsored by Black Efforts for Soul in Television and the Urban Communications Group. The meeting's goal was to increase racial minorities' awareness of the medium, and the invited participants included representatives from HUD-sponsored community development corporations (such as the Bedford-Stuyvesant Restoration Corporation), the FCC, the Model Cities Planning Council in Dayton, and the National Cable Television Association. FCC Commissioner Nicholas Johnson, Watts Communications Bureau organizer Don Bushnell, and Ralph Lee Smith also participated. Although RAND staff were not on the agenda at the 1971 conference on cable and minorities, there is evidence that these think tanks had good relations. RAND and Urban Institute documents frequently acknowledged the assistance of colleagues at the other institution.[63]

To publicize the goals of the 1971 meeting, Charles Tate, a senior researcher, released the report *Cable in the Cities.* This document, and his follow-up writings, cited findings from RAND and MITRE to call for political action on cable policy and planning. "Cable television may be the last communications frontier for the oppressed," Tate wrote. Yet, he went on, too few activists recognized the revolutionary potential of this technology. Tate favored minority ownership of local cable systems to combat the kinds of monopolies on radio and broadcast television held by "white entrepreneurs."[64] At its publication in 1971, Tate's report was the most comprehensive statement to date of how cable and minority concerns might intersect. It was also one of the few to warn of the negative consequences of surveillance, increased depersonalization of municipal services, and the role of television as a socialization instrument. It seemed to be a direct follow-up to the Kerner Commission's call to increase minority presence on and behind the television screen.

In its report, the Kerner Commission had recommended creating an institute for urban communications. It was at the Urban Institute that this center came close to becoming a reality. Shortly after the report and order, the Markle Foundation offered $500,000 to create the National Cable Television Information Center (CTIC) within the Urban Institute—a home less for research on cable television policy, as at RAND, and more a site for putting principles into practice. W. Bowman (Bill) Cutter, a former Harvard graduate, Rhodes Scholar, and graduate of Princeton's Woodrow Wilson School (he had also been initiated into the club of defense intellectuals through studies on Colombia that he conducted for the Ford Foundation) signed on as executive director.

The Cable Television Information Center (occasionally referred to as the National Cable Information Center) met immediate success. According to Cutter in October 1972, after less than a year in operation the center was already working with more than four hundred cities interested in securing cable franchises.[65] One component of this success was that many cities already had developed working relationships with Urban Institute staff on other matters during the institute's first four years of operations. Another component of this success was that the center did not charge public officials for advisory consultations with its staff of six regional directors.

The center had no single design that it was pushing as the model for all urban cable systems; each regional director had great flexibility in making recommendations. Indeed, as part of its mission, the Center eschewed any sort of model ordinance, and its official reports repeatedly cite the value of assessing local needs. As staff member Stanley Gerandasy (a former consulting engineer to the Ford Foundation) explained, system design and programming ideally would reflect each individual city's needs.[66] Center publications—titles such as *Cable Television: Options for Jacksonville* (1972) and *Planning Interconnective Systems: Options for the Twin Cities Metropolitan Area* (1974)—reflect this emphasis on local concerns. In these individualized studies, CTIC staff analyzed multiple permutations of cable systems (comparing, for example, local versus regional), presented information on state and local laws related to decision making, prepared cost-benefit analyses to quantify the risks of each option, and offered recommendations on organizing the decision-making process.

In some cities, CTIC staff played advisory roles beyond the franchise decision-making process. When William Schaefer, Baltimore's mayor, appointed a committee of citizens to consider a cable franchise soon after the FCC report and order, it included CTIC's Ellen Roberts.[67] The committee recommended that the city create an office of telecommunications on the model of New York City's, which it did in September 1973. This office survives today as the Mayor's Office of Cable and Communications. To increase citizen awareness of the medium, Baltimore launched a program to introduce citizens to cable at the 1974 city fair, and invited CTIC's Vic Nicholson to design the demonstration. That same year, CTIC's Bill Cutter received an NSF grant to test how well cable delivered social services to the elderly and engaged citizen participation in Peoria, Illinois.[68]

The Urban Institute experienced its own version of the larger cable story. As time went on, the goals of social service delivery, community development, and serving minority interests became less and less central to most city officials' interest in cable. As early as October 1972, at a conference cosponsored with MITRE, William Gorham, the institute's president, expressed a more moderate view about the future of cable as an urban medium. Addressing his remarks to those who got "tears in their eyes" when they envisioned cable's possible impact on social problems, he pointed out that the original aim of cable had been to "move pictures around mountains."[69] The cable industry would likely influence people's lives, he told the large audience; but, he cautioned, suggestions that it would eliminate poverty and other urban problems were likely misguided. Gorham's reminder that the cable technology originally had been developed for technical purposes and his caution about the difficulties adapting it to social purposes was an unusual admission at a conference of cable enthusiasts. The CTIC continued to offer cable advice, but the connection to problem solving eventually fell away. Eventually, the center split from the Urban Institute, becoming an independent operation and moving to Virginia in the 1980s.

Historians of technology are familiar with the idea that certain aspects of a technology's design, a response to specific historical contingencies, often persist long after they are technically or socially necessary. For example, the QWERTY lettering on English-language computer keyboards harks back to an era of mechanical typewriters, whose keys jammed if typists worked too quickly. Generations ago, the QWERTY design, by separating commonly used letters, slowed down typists to eliminate this problem. Its presence on computer keyboards today reflects the powerful sway of social conventions on science and technology. Public-access channels should be understood similarly, as artifacts of a specific historical moment. In an era when political leaders wanted to encourage peaceful interaction of their citizenry and to deter public violence, community channels seemed to offer a means to do so, a noncontroversial approach to integration. Cable operators also championed public access early on, likely as a ploy to gain support when they were seen as undermining broadcasters by offering a competing service.[70] Yet almost as soon as these channels were mandated by law, their relevance was outmoded. Visions of cable inspired by cold war social science did not translate well into a market environment.

Despite great investment by defense intellectuals and the think tank community, their participation did not have significant effects on the kinds of programming that would dominate the medium for the coming decades. In the period immediately following the FCC report and order, local pressures to obtain franchises persisted, and city officials invited the defense intellectuals to offer their advice. Dayton asked for assistance from RAND. Jacksonville, Minneapolis, and Saint Paul called in the Cable Television Information Center.[71] Cincinnati invited a variety of consultants from MITRE (Ralph Lee Smith) and from other city projects (Herbert Dordick) and even sent its Cable Television Task Force to survey the situation in New York City. Yet by the early-1970s, visions of wired cities from the mid-1960s already were out of step with many city planners' and managers' everyday concerns about efficient delivery of municipal services. A different financial and political climate in many urban areas had emerged during the freeze, one that siphoned support for uses more complex than entertainment and diminished interest in the medium as a civic technology. City governments never had the same access to financial resources as the military, and this was especially true during the financial crises of the early 1970s. Without guaranteed multiyear federal or foundation support, cities could not go it alone and invest in creative pilot projects. A survey of the state of the nation's cable systems in the mid-1970s found that in only four cities had government channels been programmed—in Tulsa, Oklahoma; Madison, Wisconsin; Somerville, Massachusetts, and Rockford, Illinois.[72]

As Herman Kahn had predicted in 1972, plans for cable imagined during the freeze were unable to account for the ways in which much of the "real world" operates—not on advance planning but on last-minute problem solving—and how changing political and economic circumstances in American cities made experts' models, and even the public policies they had inspired, difficult to apply. In an analysis of cable developments to 1975, Kas Kalba offered the view that all the planning might have backfired. Kalba wrote that the extensive investment in franchising models ironically may have had contradictory or regressive effects. Models were developed during the freeze when the assumption was that cities would jump into franchising and that cities needed guidelines to ensure that the public interest was served. Yet many defense intellectuals did not anticipate that "a city's decisions on whether to franchise and when to franchise would turn out to be as important as how to franchise."[73] Indeed, some cities such as Boston and Kansas City declared that they would have nothing to do with cable, explicitly backing away from franchising decisions. By late-1973, many cable companies began to pull out of large city markets.[74] Cities with cable already in place, such as New York, experienced their own version of these events. Thus, except for a few sites where broadcast television reception was poor, making cable a necessary answer to a technical problem, there did not seem to be adequate reasons for cities to sponsor and for consumers to buy the service or even invest in a pilot project.

By the late-1970s and early-1980s, the fate of cable was sealed. Cable operators assumed financial responsibility for wiring the nation's cities, and entertainment and profit, not problem solving or investing in public-access centers, were foremost in their minds. Reregulation during the 1970s meant that the FCC and courts revised cable regulations repeatedly, in ways that favored corporate interests. For example, a 1976 lawsuit from Midwest Video over public-access requirements for communities of more than thirty-five hundred residents led to a 1979 Supreme Court ruling that the FCC had overstepped its bounds. The Court decided that local areas could choose whether to mandate public-access channels. In this climate, cable companies continued to fund the development of basic infrastructure, but shied away from developing two-way communications or sophisticated production facilities for public access.[75]

By 1984, 71 percent of the country had access to cable. The Cable Communications Policy Act that year further preempted local regulations. The act reflected Congress's recognition that the technological infrastructure had moved away from its early role as a civic technology. According to one interpreter, the 1984 act "was essentially written by the cable industry," and the president of the National Cable Television Association went so far as to frame a copy and display it "as a sort of hunter's trophy."[76] Thus, even though many franchises of the late-1970s and early-1980s were based on municipal boundaries, the cable systems operating in this period were not municipal information services, but rather were entertainment systems that provided more channels than regular broadcast television. Cable operators gave back a small percentage of their profits to cities in exchange for access to a particular municipal population of subscribers, but goals for enhancing social stability, community development, and public access in poor areas, or any sense that the medium was a public good like a utility, were no longer in the foreground.

The impact of cable's first decade as an urban technology, like the impacts of systems analysis, computer simulations, space age management, and satellite reconnaissance on actual decision making and quality of urban life, was far less substantial than had been hoped. Key differences between the organizational contexts of military and civilian settings had impeded easy transfer of these tools. Similarly, despite the alliances forged between grassroots urban activists and defense intellectuals around cable communications, efforts to incorporate concerns about domestic security into this new urban infrastructure had few successes in the implementation phase. Key differences between the funding climate for large-scale military projects and their civilian analogs help to explain why cities ceded responsibility for developing municipal cable infrastructure to private companies. Financial factors also help to explain how entertainment programming came to overshadow social welfare programming. By the late-1970s, the cable medium had not realized its projected goals for revolutionizing communications among citizens or between citizens and government. And America's inner cities, with their concentrations of low-income residents, were more often last, rather than first, to be wired.[77]

Yet the defense intellectuals' failure to anticipate obstacles to delivering on their promises did not diminish their public stature. As advisers on prominent cable task forces and as directors of sponsored cable research, they used interest in the medium as a means to nurture their own social networks. More than two decades later, the MITRE Corporation maintains a civil communications division for outreach to local governments. The Urban Institute's board still includes individuals with defense and international security expertise (among them, John Deutsch, the chemist and former CIA director). The Ford Foundation continues to fund urban studies, media studies, and security studies. The Markle Foundation sponsors projects ranging from the Task Force on Security in the Information Age (together with the Brookings Institution and the Center for Strategic and International Studies) to the Global Digital Opportunity Initiative (together with the UN Development Program) to deliver information technology to underserved populations across the globe. And at the start of the twenty-first century, in an era when observers again characterize the nation's cities as "information networks," Lloyd Morrissett has found fame as the man who coined the phrase *digital divide* to refer to the persistent belief that isolation from communications channels underpins social and economic inequalities in the information age.[78] It is the longevity of these human networks, as much as the failure to create urban cable networks, that characterizes the lasting significance of this episode for American urban history.

Cable television has continued to thrive in a different form from what had been predicted by the defense research community. In some areas, municipal channels and community access remain, although programs of local origin are now a fringe element, with limited viewership, rather than the fundamental rationale for cable systems.[79] Most of all, the alliance between cable technology and entertainment broadcasting has been clearly symbolized by a name change. The innovation referred to throughout the 1960s and into the 1970s as cable communications has dropped its maiden name to become cable television, in a marriage that has lasted to the present day.

Conclusion

Histories of twentieth-century science and social science in the United States are filled with discussions about how large-scale investments in defense research and development have changed the practices of many fields, from communication research to physics to psychology. Scholars find links to the needs of U.S. military and intelligence agencies in the founding of prestigious university centers and laboratories, in individual research careers, and in financial support for the study of specific problems. Sometimes, for example, in the work of MIT's Radiation Laboratory (Rad Lab) or Caltech's Jet Propulsion Laboratory (JPL), these connections were made public. At other times, for example in the Manhattan Project or the CIA-backed Center for International Studies at MIT, the relationships were kept covert.[1]

This book has brought to light a parallel transformation in American urban planning and management during the cold war. As academic disciplines taught in schools of design and public administration, and as professional practices in city agencies, urban planning and management metamorphosed in interaction with the nation's security establishment. The life stories of individuals and institutions presented here suggest how attending to connections between cold war planning and management and urban planning and management forces us to rethink the recent history of urban research and to reinterpret the recent history of U.S. cities.

Visionaries such as Norbert Wiener and Paul Baran, central actors in the history of communication and information technology, may not previously have occupied a significant place in accounts of America's urban development in the decades since World War II. They are, however, paradigmatic examples of a new type of urban expert who emerged in the postwar period. From their scientific work on defense projects, Wiener and Baran went on to propose that communication and information theories and technologies be applied to urban problem solving. Maintaining security in cities—whether the threats be external or internal—was a central theme of their arguments. While both men's contributions to urban planning and management remained theoretical, their ideas were adopted and implemented by individuals and organizations who believed that the urban applications of defense innovations might help to save American cities.

The belief that the United States could address its urban problems by continuing to channel funding toward the defense and aerospace communities was possible only in a society where, to quote Senator William Fulbright from 1969, "not only to the Strangeloves . . . but to millions of honest, decent Americans whose primary concern is with nothing more than earning a decent living for their families . . . violence has become the nation's leading industry."[2] Despite President Eisenhower's 1961 warning, the expansion of the military-industrial complex continued. Outrage from black activists about the ongoing "war" in America's inner cities may be overstated, yet these claims captured a fundamental reality—that for several decades, a continuing source of inspiration for U.S. city planners and managers lay in the nation's preparations for war.[3]

Mobilization and Retreat

One of the hallmarks of U.S. approaches to warfare from World War II onward is how the nation's defense and space agencies successfully recruited the brightest minds of the era to their service. The task of these intellectuals was not frontline combat, but research for national service. As a result, wartime priorities steered individual and institutional attention toward specific research topics and methods. Many cold war–era studies in the sciences and social sciences—including some whose motivations appeared to be "basic research"—in fact reflected the needs or followed the leads of defense and space-agency patrons.[4]

The intellectual history of urban research in several fields fit this pattern. Project East River enlisted planning professors from leading U.S. academic institutions to forecast potential urban effects of an atomic attack, and in their later research many of the participants continued to investigate urban decentralization. Defense mapping efforts during the cold war gave university geographers access to remote sensing instrumentation, and later these scholars led the charge to make such instruments more central to urban analysis. And years of geographic information system development at Harvard's Graduate School of Design—including those systems whose military applications were not immediately evident—were made possible thanks to funding from the Office of Naval Research.

This patronage-driven reorientation of academic leaders in urban research toward defense-related topics was an important turn of events in the history of city planning and management. Yet even more significant for the history of urban research—and, in turn, for the history of American cities—is how leaders in defense and aerospace research and development found their own institutional attention turned toward urban topics. Think tanks, the aerospace industry, and other defense research installations have not been central actors in previous histories of urban planning and management.[5] Yet from the 1950s onward, these institutions served as vital sources of inspiration for new directions in urban analysis—also, in part, a result of pressures from above.

During the 1950s and 1960s, America's defense and intelligence agencies expanded their vision of security strategy to embrace an interest in the socioeconomic and demographic dimensions of enemy populations alongside an earlier interest in obtaining knowledge about physical infrastructure and weapons capabilities. This new priority in national security studies, which helped to shape the course of urban research at universities, also affected operations at institutions such as RAND, Lockheed, and JPL. Techniques from systems analysis to airphoto interpretation came to reflect new attention to urban social data, and technologies from computer simulations to geographic information systems helped make sense of these data for war planning and management.

Thus, years before President Johnson called in an army of defense intellectuals to fight the War on Poverty in the nation's cities, the research and development arm of the U.S. security establishment already had taken a strong interest in urban analysis. With the escalation of urban crisis at home and an increase in federal spending on domestic issues during the 1960s, defense intellectuals from universities, think thanks, and aerospace companies were well positioned for entry into urban markets. Confidence in the wide applicability of their knowledge and simultaneous anxiety about institutional longevity should preparations for war de-escalate together laid the groundwork for years of commitment to the problems of the nation's cities. Reframing the urban crisis as a national security crisis helped to transform urban problems into strategic challenges for which their expertise was uniquely suited.

Urban planning and management, like war planning and management, depend on research and development for new ideas and approaches. But above all, they are applied endeavors. Fortunately for the defense intellectuals, the notion that ideas and innovations originally developed to defend against foreign enemies of the United States might wage a successful campaign against domestic urban blight, chaos, and unrest was one that many other constituencies wanted desperately to believe. Chief among them were the nation's city planners and managers. When the defense intellectuals recast themselves as experts on urban problem solving, city administrators were seduced. New York City and RAND, Pasadena and Space-General, Los Angeles and JPL, San Jose and Lockheed, Dayton and RAND. . . . That administrators in so many cities invited these institutions to become fellow soldiers in the attack on urban problems was the critical step in remaking the defense intellectuals' legacy from one of intellectual influences on urban research to one that included influences on American city operations.

Influences do not equal successes, however, and the legacy of these collaborations is decidedly mixed. Certainly, when it came to mobilizing resources for the problem solvers, these efforts succeeded brilliantly. Think tanks and aerospace companies found contracts as consultants to city governments, and ammunition to defuse hostility from critics who argued that federal defense spending was diverting money from the nation's domestic problems. NASA found that management techniques and reconnaissance technologies developed for space had spin-offs on Earth that could generate support for expensive future missions. Social scientists with expertise applying communication theories and technologies in development programs to control violence and insurgency outside U.S. borders found that this approach could be applied to maintain their position as advisers to government by shaping the uses of cable communications for urban development on the home front. And mayors and city managers found political clout from their new technocratic approaches to attacking urban blight.

When it came to delivering on their promises of urban problem solving, the mostly cordial collaborations documented throughout this book did not give the transfer-of-innovation process high marks. New city agencies such as the Los Angeles Community Analysis Bureau and the New York City RAND Institute were created to improve city management and residents' social welfare on the model of defense research organizations. While their experiments in community renewal and housing planning using military management tools streamlined some decision-making processes, few changes in to the day-to-day character of urban life were observable. NASA promoted satellite surveys as a technically advanced method for cities to gather comprehensive information to improve decision making, especially when integrated into geographic information systems. Instead, aerial photography, a military innovation already widely in use, maintained its position as the standard medium for physical and social planning. A vision of cable communications as a municipal information utility sparked the interest of defense intellectuals and city leaders as a means to offer social services to, and enhance community participation among, disenfranchised urban residents. Yet cable infrastructure and programming moved in a direction entirely different from what these reformers had envisioned.

The mobilization of resources to wage war on urban problems was thus not the same as winning the battle, and as the failure of efforts to adapt defense and aerospace ideas and innovations and expertise to the urban context became more widely recognized, some of the forces supporting these collaborations began to retreat. RAND, controversial for the amount of money paid to its consultants in the absence of visible results from their work, dismantled its New York City branch in 1975. Operation Breakthrough, which influenced the production of some housing components, hung on through a change of administration at HUD but was discontinued soon after. And experiments using cable communications for community development in several cities during the 1970s, dependent on continued foundation and federal sponsorships, largely were suspended by the end of the decade.

Post–Cold War Echoes

The stories recounted here have underscored that, despite all of the brain-power, cross-sector cooperation, and federal support, transfers of ideas and innovations from military to urban applications usually missed their target. Perhaps it is therefore unsurprising that most accounts of U.S. city planning and management have devoted little attention to this topic. Yet only by restoring connections between military and urban planning and management to a more prominent place in American urban history can scholars and practitioners begin to appreciate how deeply cold war techniques, technologies, and institutions have continued to shape urban analyses and city operations to the present day.

The late-1960s marked the peak of the defense intellectuals' public participation in city planning and management. Yet neither the disappearance of a sense of urban crisis from the political scene nor disappointments with the transfer of specific innovations permanently tarnished the public profile of the experts or their methods of analysis. Influences from these experiments in urban problem solving lingered long beyond their formal conclusions, shaping the "image of the city" for the post–cold war information age.

Urban analysts at universities and think tanks have continued to depend on computer simulations for planning and policy decision making. City planners and managers have continued to work with technologies whose genesis was military operations, ranging from command and control systems for police dispatch to GIS for monitoring of public health to urban simulations for participatory community development. Aerospace firms including Lockheed Martin and Northrop Grumman have continued to maintain divisions for bringing innovations in information management to municipal clients. Government and academic leaders have continued to conceptualize inequality as a problem of communication, as efforts to address the digital divide bank on the belief that the key to economic development and social integration lies in wider access to communication technology—this time around, the internet.[6]

Connections to their cold war roots may be long forgotten, yet each of these efforts exemplifies how the intellectual, technical, and institutional weaponry used to fight the cold war have inspired another generation of approaches to analyzing, planning, and managing cities and their problems.[7] From the cold war's end in the late-1980s, through a decade of economic expansion for the nation's technology sector in the 1990s, the necessity of adopting information and communication technology—rather than the necessity of maintaining urban security—provided the chief rationale for continued engagement with the urban applications of ideas and innovations first developed to meet cold war military needs.[8]

Urban security returned to a prominent position on the national political agenda following the September 11, 2001, terrorist attacks. Questions reminiscent of the 1940s resurfaced as political leaders, military officials, mayors, and city dwellers asked: How can concentrated metropolises reduce their vulnerability to future attacks? Is it safer to live and work in the suburbs? When America's military-industrial-academic complex first mobilized for a War on Terrorism, civil defense for cities lay at the heart of its domestic program.[9]

The long-term influences of this war on how the nation plans and manages its cities must await future inquiry. For now, the aftermath of September 11 has brought to the national consciousness a recognition that defense experts have an essential role to play in shaping future operations in U.S. cities. Already, these experts have left indelible marks, for better and for worse, on the nation's urban past.

Notes

Introduction

1. National League of Cities, *City of Man: Proceedings of the Forty-third Annual Congress of Cities, Las Vegas, Nevada* (Washington, D.C.: National League of Cities, 1966), 4.

2. National League of Cities (1966), 4–6.

3. National League of Cities (1966), 7; Donald Janson, "Webb Backs Cost of Space Program: Rebuts Attack on Priority at Parley of City Officials," *New York Times*, December 6, 1966, 34.

4. The term *aerospace* dates to 1958, but it was used retroactively to refer to innovations developed earlier.

5. Adam Yarmolinksy, *The Military Establishment: Its Impacts on American Society* (New York: Harper & Row, 1971).

6. Yolanda Ward, "Spatial Deconcentration," www.interactivist.net/housing/spatial_d_1.html; Samuel Yette, *The Choice: The Issue of Black Survival in America* (New York: Putnam, 1971). I use the term *defense intellectual* throughout this book as shorthand for a category of civilian experts who participated in defense planning; the "intellectuals" are to be differentiated from men with distinguished military careers who moved into civilian government. C. Wright Mills, *The Power Elite* (New York: Oxford University Press, 1950) is a classic study of this latter phenomenon in the American context. As with the categories "scientist" and "politician," the civilian defense intellectuals were not an entirely homogenous group.

7. The military shorthand for this phrase is C4ISR, which postdates the time period discussed in this book but perfectly captures the range of defense innovations exported to city governments.

8. Edward Relph, *The Modern Urban Landscape* (Baltimore: Johns Hopkins University Press, 1987); Jon Teaford, *The Rough Road to Renaissance: Urban Revitalization in America, 1940–1985* (Baltimore: Johns Hopkins University Press, 1990); Thomas Sugrue, *Origins of the Urban Crisis: Race and Inequality in Postwar Detroit* (Princeton, N.J.: Princeton University Press, 1996).

9. Thomas Hanchett, "Federal Incentives and the Growth of Local Planning, 1941–1948," *Journal of the American Planning Association* 60, no. 2 (1994): 197–208; Donald Albrecht, ed., *World War II and the American Dream: How Wartime Building Changed a Nation* (Cambridge: MIT Press, 1995); Tom Lewis, *Divided Highways* (New York: Viking Penguin, 1998); Roger Lotchin, *Fortress California, 1910–1961* (New York: Oxford University Press, 1992); Ann Markusen, Peter Hall, Scott Campbell, and Sabrina Detrick, *The Rise of the Gunbelt* (New York: Oxford University Press, 1991).

10. The term *technique* refers to methods of analysis—for example, systems analysis and photointerpretation—as well as methods dictating practices including space age management and development theory.

11. "Military-industrial-academic complex" is Stuart Leslie's (1993) updating of President Dwight D. Eisenhower's (1961) "military-industrial complex." Stuart Leslie, *The Cold War and American Science: The Military-Industrial-Academic Complex at MIT and Stanford* (New York: Columbia University Press, 1993); Dwight Eisenhower, "Farewell Address": http://mcadams.posc.mu.edu/ike.htm

12. John Staudenmaier, *Technology's Storytellers* (Cambridge: MIT Press, 1985); Richard Rhodes, *The Making of the Atomic Bomb* (New York: Simon & Schuster, 1986); Peter Galison and Bruce Hevly, eds., *Big Science: The Growth of Large-scale Research* (Stanford, Calif.: Stanford University Press, 1992); Daniel Kevles, *The Physicists: The History of a Scientific Community in Modern America* (New York: Knopf, 1977); David Noble, *Forces of Production: A Social History of Industrial Automation* (New York: Knopf, 1984); Les Levidow and Kevin Robins, eds., *Cyborg Worlds: The Military Information Society* (London: Free Association Books, 1989).

13. In some cases this has led to a distortion of urban analysis as scholars project onto all cities the experience of these two larger-than-life urban centers; see my discussion, re: Los Angeles, in Jennifer Light, "From City Space to Cyberspace," in Phil Crang, Mike Crang, and Jon May, eds., *Virtual Geographies* (London: Routledge, 1999). In a similar vein, many of the defense and security experts described throughout this book, enthusiastic promoters of the transfer of innovations from military and aerospace to urban settings, were unable fully to appreciate the range of environments encompassed by these terms and how local conditions might alter the outcomes of technology transfer.

14. Although this book extends studies of the cold war on American civilian culture, it emphatically is not an argument about the "militarization of everyday life." City planners and managers across the United States adopted the products of military research and development, but the effects of these technologies on the average urban dweller's existence were negligible. The urban effects of information and communication technologies discussed throughout the book were largely invisible to the public since they occurred at the organizational level; in other words, rather than shaping, for example, the physical design of the Los Angeles City Hall, cybernetics and computer simulations affected the city's urban management practices.

15. *Power elite* is C. Wright Mills's term. A critical early appraisal of the military-industrial complex that predates the term itself is in C. Wright Mills, *The Causes of World War Three* (New York: Simon & Schuster, 1958), which extended his ideas from *The Power Elite* (1956) to critique the mobilization for total war that he predicted would precipitate a nuclear holocaust.

16. *Cybercities* is a term popularized by Christine Boyer: Boyer, *Cybercities* (New York: Princeton Architectural Press, 1996).

ONE: Planning for the Atomic Age

1. Louis Wirth, "Does the Atomic Bomb Doom the Modern City?" in *Proceedings of the Twenty-second Annual Conference of the American Municipal Association* (Chicago: American Municipal Association, 1945), 30–33. The American Municipal Association was precursor to the National League of Cities.

2. Constance Perin, *Everything in Its Place: Social Order and Land Use in America* (Princeton, N.J.: Princeton University Press, 1977); Kenneth Jackson, *Crabgrass Frontier: The Suburbanization of the United States* (New York: Oxford University Press, 1985); Joel Garreau, *Edge City: Life on the New Frontier* (New York: Anchor Books, 1991); Michael Sorkin, ed., *Variations on a Theme Park: The New American City and the End of Public Space* (New York: Hill & Wang, 1992); Edward Relph, *The Modern Urban Landscape* (Baltimore: Johns Hopkins University Press, 1987).

3. Jay Winter and Jean-Louis Robert, *Capital Cities at War: Paris, London, Berlin, 1914–1919* (New York: Cambridge, 1997); Jeffrey Diefendorf, *In the Wake of War: The Reconstruction of German Cities after World War II* (New York: Oxford University Press, 1993); Charles Whiting, *Britain under Fire: The Bombing of Britain's Cities, 1940–1945* (London: Century, 1986); Gregory Ashworth, *War and the City* (New York: Routledge, 1991). Sources on the relationship between national security needs and U.S. cities are discussed in the note on sources, following the notes.

4. Coleman Woodbury, ed., *The Future of Cities and Urban Redevelopment* (Chicago: University of Chicago Press, 1953), 166–67. Despite their presence in the popular imagination, the interstate highway network and the fallout shelter program, often cited as the best-known examples of the ways that defense needs shaped the physical environment in the United States, are not as relevant here as dispersal. Discussions about highways and shelters did not produce the same kinds of social networks as dispersal; nor did discussions engage the question of the relationship between military and social welfare planning that is the focus of this book.

5. The Franck Report can be found at www.dannen.com/decision/franck. The Los Alamos document is described in Woodbury (1953), 169.

6. Jacob Marschak, Edward Teller, and Lawrence Klein, "Dispersal of Cities and Industries," *Bulletin of the Atomic Scientists* (hereafter, *BAS*) 1, no. 9 (1946): 13–15, 20; *Atomic Bomb Explosions: Effects on an American City*, prepared by Ralph Lapp, Research and Development Division, War Department (Washington, D.C.: U.S. War Department, 1947); U.S. Atomic Energy Commission, *The City of Washington and an Atomic Bomb Attack* (Washington, D.C.: Atomic Energy Commission, November 17, 1949), VF NAC 3810.5g27 Wash., Harvard University Graduate School of Design Library Vertical Files Collection (hereafter, VF).

7. Tracy Augur, "Planning Cities for the Atomic Age," paper prepared for the New York meeting of the American Institute of Planners, May 5, 1946, 3, VF NAC 3810.

8. Greg Herken, *The Winning Weapon* (New York: Alfred Knopf, 1980), 211–12; Harold R. Bull, "Gen. Bull Warns Mayors on 'A' Bomb," February 18, 1948, 1, VF NAC 3810; and Harold R. Bull, "What an Atomic Bomb Might Do to Your City," *American City* 63 (March 1948): 83; Ernest Oppenheimer, "The Challenge of Our Time," *BAS* 3, no. 12 (1947): 373; Leslie Groves, "Dispersal of Industry in the Atomic Age," *Mechanical Engineering* 77 (June 1955): 486–87; E. Despres, *Dispersal* (Santa Monica, Calif.: RAND, August 21, 1950); David McGarvey, *Intra-city Dispersal: Some Calculations of Opportunity* (Santa Monica, Calif.: RAND, October 20, 1959); H. Markowitz and Stephen Enke, *The Protection of Target Plants through Dispersal* (Santa Monica, Calif.: RAND, January 3, 1952).

9. Burnham Kelly, "The Necessity for Dispersion" *Journal of the American Institute of Planners* (hereafter, *JAIP*) 19, no. 1 (1953): 20–25, reprinted from MIT Department of Civil Engineering, *Proceedings of the Conference on Building in the Atomic Age* (Cambridge: MIT, 1952), 108–11.

10. Ansley Coale, *The Problem of Reducing Urban Vulnerability to Atomic Bombs* (Princeton, N.J.: Princeton University Press, 1947); Philip Hauser, Otis Dudley Duncan, and Beverley Duncan, *Methods of Urban Analysis* (Maxwell Air Force Base, Ala.: Maxwell Air Force Base, 1955), VF NAC 815 H; Associated Universities, *Report of the Project East River* (New York: Associated Universities, 1952); Michael Dudley, "Sprawl as Strategy: City Planners Face the Bomb," *Journal of Planning Education and Research* 21, no. 1 (2001): 52–63.

11. Tracy Augur, "The Dispersal of Cities as a Defense Measure," *BAS* 4, no. 5 (1948): 131–34; idem, *JAIP* 14, no. 3 (1948): 29–35.

12. Otto Nelson, "The Planning of American Cities in an Atomic Era," address at ASPO conference, Detroit, October 12, 1953, 5, VF NAC 3810.5 N.

13. Thomas Hanchett, "Federal Incentives and the Growth of Local Planning, 1941–1948," *Journal of the American Planning Association* 60, no. 2 (1994): 204.

14. Gershon Cooper and Roland McKean, "Is Dispersal Good Defense?" *Fortune*, November 1954, 126–27, 161, 163; Frank Zeidler, "A Mayor Looks at the Civil Defense Program," *BAS* 6 (August/September 1950): 249–51.

15. Augur, "Dispersal" (1948); Augur (May 5, 1946); Tracy Augur, "Planning Principles Applied in Wartime" *Architectural Record* 93 (January 1943): 71–82; Tracy Augur, "For Defense and Better Living," in *American City* 63, no. 6 (1948): 5.

16. Augur, "Dispersal" (1948), 29–30.

17. Augur, "Dispersal" (summer 1948), 30.

18. Augur (May 5, 1946), 4.

19. Augur (May 5, 1946), 16–17.

20. MIT Department of Civil Engineering (1952), 110.

21. Tracy Augur, "Industrial Growth in America and the Garden City," MLA thesis, Harvard University, School of Landscape Architecture, 1921.

22. Jane Jacobs, *The Death and Life of Great American Cities* (New York: Vintage, 1961).

23. Kelly (1953), 21; Marschak, Teller, and Klein (1946); Oppenheimer (1947); Amos Hawley, "Urban Dispersal and Defense," *BAS* 7, no. 10 (1951): 307–12; and Cooper and McKean (November 1954).

24. Catherine Bauer, "Redevelopment: A Misfit in the Fifties," in Woodbury (1953), 13–25.

25. MIT Department of Civil Engineering (1952), 109.

26. Robert Moses, city construction coordinator, memo to Mayor William O'Dwyer, "Construction of Civilian Defense," August 7, 1950, in O'Dwyer Papers, subject files 1946–50. box 20, folder 194, Civil Defense (3) 1950, Municipal Archives, NYC (hereafter, MANY); Robert Moses, "Defense and Local Improvements," *American City* 55, no. 10 (1940): 109.

27. Clarence Stein, "Do New Towns Provide Safety? Yes," *Progressive Architecture* 32, no. 9 (1951): 77–79; Donald Monson, "Is Dispersal Obsolete?" *BAS* 10 (December 1954): 378–83; Goodhue Livingston, "The Blight of Our Cities," *BAS* 7 (September 1951): 260–62; and Lemuel Dillenback, "Planning for Defense and City Planning," excerpts from remarks before the meeting of New York State Federation of Official Planning Boards and state conference of mayors and municipal officials, Syracuse, N.Y., June 9–10, 1942, *New York State Planning News* 5, no. 5 (1942): 1–8.

28. Bauer (1953), 24.

29. U.S. Strategic Bombing Survey, *The Effects of Atomic Bombs on Hiroshima and Na-gasaki* (Washington, D.C.: GPO, 1946); Atomic Energy Commission (1949); Federal Civil Defense Administration, *Civil Defense Urban Analysis* (Washington, D.C.: GPO, 1953).

30. Arthur Wallander and Robert Moses, *New York City Civil Defense* (New York: Office of Civil Defense, 1951); University of Maryland, Bureau of Business and Economic Research, *Baltimore and the H-bomb* (College Park: University of Maryland, College Park, 1955); State of New York Committee on Fallout Protection, *Survival in a Nuclear Attack* (Albany, N.Y.: [Committee on Fallout Protection], 1960), VF NAC 3810g27 NY; Massachusetts Committee on Public Safety, *Organization Plan of the Massachusetts Committee on Public Safety for Civilian Defense* (Boston: [Committee on Public Safety], 1941), VF NAC 3810; Gordon Young, "Civil Defense Planning by the City of Washington, D.C." (Washington, D.C.: Author, 1950), VF NAC 3805g27 Wash; Denver Planning Office, *Proposed Research to be Included in Survival Plan: Reduction of Urban Vulnerability* (Denver: [Denver Planning Office], 1956), VF NAC 3805g27; Oscar Sutermeister, *Reduction of Vulnerability in the Milwaukee Area* (Milwaukee: N.p., 1954); Los Angeles City–owned Department of Water and Power, *Industrial Dispersal for Defense in the City of Los Angeles* (Los Angeles: Department of Water and Power, 1952), VF NAC 8102g27 osA; Chicago Civil Defense Committee, *Chicago Alerts* (Chicago: Chicago Civil Defense Committee, 1951).

31. Letter to Walter Blucher, executive director of the American Society of Planning Officials, from Jerry Finkelstein, chair, NYC Planning Commission, August 4, 1950, in O'Dwyer Papers, subject files 1946–50, box 20, folder 194, Civil Defense (3) 1950, MANY.

32. Press release #62, for August 6, 1950, in O'Dwyer Papers, subject files 1946–50, box 20, folder 194, Civil Defense (3) 1950, MANY.

33. Lack of appropriation is described in Michael Dudley, "Sprawl as Strategy: City Planners Face the Bomb," *Journal of Planning Education and Research* 21, no. 1 (2001):52–63, although as news reports have suggested, two satellite outposts of civil service officials across agencies were created as an entire shadow government: Barton Gellman and Susan Schmidt, "Shadow Government Is at Work in Secret: After Attacks, Bush Ordered 100 Officials to Bunkers away from Capital to Insure Federal Survival," *Washington Post,* March 1, 2002, A01.

34. National Security Resources Board, *Critical Target Areas in Civil Defense* (Washington, D.C.: [NSRB], 1950); American Municipal Association, *The Municipal Viewpoint on Civil Defense, 1958* (Washington, D.C.: American Municipal Association, July 1958); Federal Civil Defense Administration, *The States, Counties, Cities and Civil Defense* (Washington, D.C.: GPO, 1958); Dorothy Tompkins, *Civil Defense in the States* (Berkeley: University of California Bureau of Public Administration, April 1953), VF Z-NAC 3805T.

35. Woodbury (1953); National Security Resources Board, *National Security Factors in Industrial Location* (Washington, D.C.: GPO, 1948); U.S. Office of Area Development, *National Industrial Dispersion Program* (Washington, D.C.: [Office of Area Development], 1957).

36. U.S. Office of Area Development, *Memorandum for Local Industrial Dispersion Groups* (Washington, D.C.: Office of Area Development, 1956), VF NAC 8102 US; U.S. Department of Commerce, *Industrial Dispersion Guidebook for Communities* (Washington, D.C.: GPO, 1952); U.S. Department of Commerce, Area Development Division, *Indus-*

trial Dispersion Program: Progress in Urban Areas of the United States (Washington, D.C.: [Area Development Division], February 16, 1953), VF NAC 8102 US.

37. Dudley (2001), 15; C. Benson Wigton, "Is Your City a Target?" *American City* 68, no. 2 (1953): 159–61.

38. University of Maryland, Bureau of Business and Economic Research, *Industrial Dispersal* (College Park: University of Maryland, 1956), 8, VF NAC 3807.

39. University of Maryland, Bureau of Business and Economic Research (March 1956), 6; U.S. Department of Commerce, Office of Technical Services Area Development Division, *Meeting of Industrial Dispersion Committees, June 16, 1954* (Washington, D.C.: [Area Development Division], 1954), VF NAC 8102 U; "Dispersal—Where's the Incentive?" *Iron Age* 177 (January 19, 1956): 27–29; and Marshall Wood, "Industry Must Prepare for Atomic Attack," *Harvard Business Review* 33, no. 3 (1955): 115–28.

40. Woodbury (1953), 184; Kenneth Rose, "Backyard Apocalypse: The Fallout Shelter Controversy of the 1960s," unpublished presentation, n.d., 2; Life Insurance Association of America, *General Aspects of Civil Defense* (New York: [Life Insurance Association], 1954); Office of Defense Mobilization, *Reducing Our Vulnerability to Attack* (Washington, D.C., May 20, 1957), 3, VF NAC 3805 U.

41. Cited in William Gill, "Address of William Gill, Assistant Director, Civilian Mobilization Office, NSRB before the American Institute of Planners in Washington, D.C." (May 15, 1950), VF NAC 3810.5 G.

42. U.S. Federal Civil Defense Administration, *Report on the Washington Conference of Mayors and Other Local Government Executives on National Security* (Washington, D.C.: Author, 1956), VF NAC 3805 W.

43. Federal Civil Defense Administration, *Battleground USA* (Washington, D.C.: GPO, 1957).

44. Woodbury (1953).

45. Kelly (1953), 20.

46. Richard Bolling and Lewis Anthony Dexter, "Safety from Atomic Attack," *New Leader*, November 29, 1954, n.p., VF NAC 3810.5 B; U.S. Department of Commerce, Office of Technical Services Area Development Division (1954), 8–9; Hubert Humphrey, "To Provide for the Common Defense," *BAS* 11 (September 1955): 244–46.

47. Philip Clayton, "Can We Plan for the Atomic Age?" *JAIP* 26, no. 2 (1960): 111–18.

48. "Civil Defense Developments," *BAS* 17, no. 8 (1961): 347; see also Roger Hagan, "The Myth of Civil Defense," *Cambridge* 5, no. 2 (1961): 9–14, VF NAC 3810.5 H.

49. Kenneth Rose, *One Nation Underground: The Fallout Shelter in American Culture* (New York: New York University Press, 2001). The system was reviewed under the Nixon administration and subsequently scrapped in favor of the Safeguard system, although Safeguard (unlike the Sentinel system) was not intended to protect civilians.

50. Henry Churchill, "What Kind of Cities Do We Want?" in Woodbury (1953), 45–51.

TWO: The City as a Communication System

1. Norbert Wiener, "How U.S. Cities Can Prepare for Atomic War: MIT Professors Suggest a Bold Plan to Prevent Panic and Limit Destruction," *Life Magazine* 29, no. 25 (1950): 85.

2. Wiener, "How U.S. Cities Can Prepare" (1950), 85.

3. Norbert Wiener, *The Human Use of Human Beings: Cybernetics and Society* (Boston: Houghton Mifflin, 1950), 134.

4. Kevin Lynch, *The Image of the City* (Cambridge: MIT Press, 1960).

5. Melvin Webber, ed., *Explorations into Urban Structure* (Philadelphia: University of Pennsylvania Press, 1964); Benjamin Bagdikian, *The Information Machines* (New York: Harper & Row, 1971); Richard Meier, *A Communications Theory of Urban Growth* (Cambridge: Harvard-MIT Joint Center for Urban Studies, 1962); Thomas Paine, "The City as an Information Network," *IEEE International Convention Record* (New York: IEEE, March 1966). Even Lynch (1960) described cities as sites for message transmission and human communication.

6. Richard Meier and Richard Duke, "Gaming Simulation for Urban Planning," *JAIP* 32, no. 1 (1966): 3–17.

7. Together, the three published the definitive paper on feedback: Norbert Wiener, Julian Bigelow, and Arturo Rosenblueth, "Behavior, Purpose, and Teleology," *Philosophy of Science* 10 (1943): 18–24.

8. Leland Swanson and Glenn Johnson, eds., *The Cybernetic Approach to Urban Analysis* (Los Angeles: University of Southern California Graduate Program in City and Regional Planning, 1964), 9. This description implied that applied cybernetics was not coercive; however, in its applications to cities some interpreters would beg to differ.

9. Steven Heims, *The Cybernetics Group* (Cambridge: MIT Press, 1991).

10. Suzanne Keller, "The Telephone in New (and Old) Communities," In Ithiel de Sola Pool, ed., *The Social Impact of the Telephone* (Cambridge: MIT Press, 1977), 289.

11. Katherine Hayles, *How We Became Posthuman: Virtual Bodies in Cybernetics, Literature, and Informatics* (Chicago: University of Chicago Press, 1999); Lily Kay, "How a Genetic Code Became an Information System," in Agatha Hughes and Thomas Hughes, eds., *Systems, Experts, and Computers* (Cambridge: MIT Press, 2000); Evelyn Fox Keller, *Refiguring Life: Metaphors of Twentieth-century Biology* (New York: Columbia University Press, 1995).

12. David Hounshell, "The Medium Is the Message, or How Context Matters: The RAND Corporation Builds an Economics of Innovation, 1946–1962," in Hughes and Hughes (2000).

13. Richard Barringer with Barton Whaley, "The MIT Political-military Gaming Experience," *Orbis* 9, no. 2 (1965): 437–58; Ithiel de Sola Pool, Robert Abelson, and Samuel Popkin, *Candidates, Issues and Strategies: A Computer Simulation of the 1960 and 1964 Presidential Elections* (Cambridge: MIT Press, 1965); Ithiel de Sola Pool, "The Simulmatics Project," *Public Opinion Quarterly* 25, no. 2 (1961): 167–83; Ithiel de Sola Pool, "The Kaiser, the Tsar, and the Computer: Information Processing in a Crisis," *American Behavioral Scientist* 8, no. 9 (1965): 31–39.

14. Wiener (1967 edition of Wiener, *Human Use* [1950]), 166.

15. Paul Edwards, "The World in a Machine," in Hughes and Hughes (2000); Claude Baum, *The System Builders: The Story of SDC* (Santa Monica, Calif.: SDC, 1981).

16. Stephen Enke, ed., *Defense Management* (Englewood Cliffs: Prentice Hall, 1967), preface; Stephen Johnson, "From Concurrency to Phased Planning," in Hughes and Hughes (2000).

17. Martin Van Creveld, *Command in War* (Cambridge: Harvard University Press, 1985), 241.

18. J. A. Wilson, *Geographic and Population Data for South Vietnam* (Santa Monica, Calif.: RAND, September 6, 1963); David Elliott and William Stewart, *Pacification and the Viet Cong System in Dinh Tuong, 1966–1967* (Santa Monica, Calif.: RAND, 1969), v.

19. Thomas Thayer, *War without Fronts* (Boulder, Colo.: Westview Press, 1985), 138, and chapter 13, "How Secure Was the Countryside?" These records are now located in the National Archives.

20. *Hamlet Evaluation System Handbook* (Saigon: Military Assistance Command, CORDS, June 1969).

21. Komer cited in Thayer (1985), xxii. The reports are compiled in Thayer's 12-vol. study *A Systems Analysis View of the Vietnam War* (Washington, D.C.: Office of Assistant Secretary of Defense for Systems Analysis, 1975).

22. Simulmatics Corporation, *A Population Survey in Vietnam* (Cambridge: Simulmatics, 1967); Robert Komer, "Text of Ambassador Komer's News Conference on the Hamlet Evaluation System," unpublished, 1967–68; James Whittaker, "Psychological Warfare in Vietnam," *Political Psychology* 18, no. 1 (1997): 165–79.

23. Erwin R. Brigham, *Pacification Measurement in Vietnam: The Hamlet Evaluation System*, prepared for the SEATO Internal Security Seminar, Manila, June 3–10, 1968 (Vietnam: CORDS, 1968).

24. Cited in Peggy Clifford, "RAND and the City, Part 3," *Santa Monica Mirror* 1, no. 21 (1999): n.p.: www.smmirror.com. These five of the nine original files (with other related files) are available in the National Archives Center for Electronic Records.

25. John Raser, *Simulation and Society* (Boston: Allyn & Bacon, 1969), 48; Enke (1967); Hounshell (2001); and David Jardini, "Out of the Blue Yonder: The RAND Corporation's Diversification into Social Welfare Research, 1946–1968," Ph.D. dissertation, Carnegie Mellon University, 1996.

26. William Lucianovic, "Analytic and Experiential Information: Gaming and the Man-machine Interface," in *Papers from the Annual Conference of the Urban and Regional Information Systems Association* (Akron, Ohio: URISA, 1970), 335.

27. C. West Churchman, in American Academy of Political and Social Science (hereafter, AAPSS), *Governing Urban Society: New Scientific Approaches* (Philadelphia: AAPSS, 1967), 37.

28. Cited in Henry Rowen, *Some Futures of Operations Research, Address to the 22nd Military Operations Research Symposium, US Naval Postgraduate School, 11 December 1968* (Santa Monica, Calif.: RAND, 1968), 14.

29. Wiener, *Human Use* (1950), 24.

30. Herbert Simon was among the other key figures in the history of cybernetics who saw early connections between cybernetic theory and city operations, likely a result of his experience working at the International City Managers' Association in 1936, after college.

31. M. C. Branch, "Planning and Operations Research," *JAIP* 23, no. 4 (1957): 168–75 (although he used the term *operations researchers*).

32. Ira Lowry, "A Short Course in Model Design," *JAIP* 31, no. 2 (1965): 158–66.

33. Darwin Stuart, *Information Systems in Urban Planning: A Review* (Chicago: American Society of Planning Officials, 1970), 2.

34. Ralph Gakenheimer, "Process Planning: Symposium on Programming and the New Urban Planning," *JAIP* 31, no. 4 (1965): 282.

35. Stanford Optner, "Systems Analysis as a Planning Tool," in *Proceedings of the 1960 American Institute of Planners Conference* (Philadelphia: American Institute of Planners, 1960): 138–45.

36. Although she did not use the term *cybernetics,* in her final chapter Jacobs praised the work of Warren Weaver, a key supporter of cybernetics.

37. Peter Langer, "Sociology—Four Images of Organized Diversity: Bazaar, Jungle, Organism, and Machine," in Lloyd Rodwin and Robert Hollister, eds., *Cities of the Mind: Images and Themes of the City in the Social Sciences* (New York: Plenum, 1984), 97–118; John Jordan, *Machine-age Ideology: Social Engineering and American Liberalism, 1911–1939* (Chapel Hill: University of North Carolina Press, 1994).

38. Meier (1962). Meier, a communications professor at the University of Michigan and also an affiliate of the MIT-Harvard Joint Center for Urban Studies, had trained as a nuclear chemist and worked for the Atomic Energy Commission before moving to the field of planning. My thanks to Jonathan Mason for bringing this to my attention.

39. Swanson and Johnson (1964), 97.

40. Swanson and Johnson (1964), 97.

41. *State of the City: Los Angeles, 1970.* (Los Angeles: CAB, 1970), vol. 4, appendix B: 79.

42. Garry Brewer, *Politicians, Bureaucrats, and the Consultant: A Critique of Urban Problem Solving* (New York: Basic Books, 1973).

43. Samuel Bernstein, "Urban Analysis and the Computer: A Strategic Approach for Urban Policy Analysis," in *Proceedings of the Association for Computing Machinery Annual Conference on the Application of Computers to the Problems of Urban Society* (New York: ACM, 1971), 15.

44. Califano, cited in Abe Gottlieb, "The Computer and the Job Undone," in *Proceedings of the Association for Computing Machinery Conference on the Application of Computers to the Problems of Urban Society* (New York: ACM, 1970), 4.

45. Lowdon Wingo, "Urban Renewal: A Strategy for Information and Analysis," *JAIP* 32, no. 3 (1966): 143–54; and Robert Wood, in AAPSS (1967).

46. Nathan Grundstein and William Kehl, *The Pittsburgh Community Model Game: A Proposal to Establish a Community Model Game with a Computer* (Pittsburgh: University of Pittsburgh, 1959); Swanson and Johnson (1964); Allan Feldt, "Operational Gaming in Planning Education," *JAIP* 32, no. 1 (1966): 17–23; *Rand New York Newsletter,* no. 3 (1968), RAND Corp. Archives (hereafter, RAND).

47. Daniel P. Moynihan, *Maximum Feasible Misunderstanding: Community Action in the War on Poverty* (New York: Free Press, 1969), 199. PPBS was one of several related tools that included the Program Evaluation Review Technique (PERT), an offshoot of the U.S. Navy's Polaris missile development.

48. Henry Rowen, telephone conversation with author, January 11, 2002.

49. Wood (1967), 179.

50. *Design Requirements for the Data and Systems Support Essential to an Urban Blight Systems Analysis: A Report* (Los Angeles: CAB, 1970), 3–4; Volta Torrey, *Science and the City* (Washington, D.C.: GPO, 1967); AAPSS (1967).

51. Kenneth Kraemer, "USAC: An Evolving Intergovernmental Mechanism for Urban Information Systems Development," in *Papers from the Annual Conference of the Urban and Regional Information Systems Association* (Akron, Ohio: URISA, 1970), 63–79;

Robert Wegner, "The Metropolitan Data Center Projects," *Public Administration Review* 22, no. 3 (September 1962): 145–46; Lawrence Williams, "The Urban Observatory Approach," *Urban Affairs Quarterly* 8 (September 1972): 5–20; "The HUD Urban Systems Engineering Demonstration Program: An Overview," in *Proceedings of the Hawaii International Conference on System Sciences* (Honolulu: University of Hawaii Press, 1973).

52. L. Lessing, "Systems Engineering Invades the City," *Fortune,* January 1968, 154–57, 217–21; National League of Cities, *City of Man: Proceedings of the Forty-third Annual Congress of Cities, Las Vegas Nevada* (Washington, D.C.: National League of Cities, 1966), 43. Reports from the NYC Community Renewal Program also made occasional reference to "programming renewal."

THREE: Cybernetics and Urban Renewal

1. James Beshers, ed., *Computer Methods in the Analysis of Large-scale Social Systems* (Cambridge: MIT Press, 1965), 219.

2. Martin Anderson, *The Federal Bulldozer* (Cambridge: MIT Press, 1964).

3. *Design Requirements for the Data and Systems Support Essential to an Urban Blight Systems Analysis; A Report* (1970), 5.

4. Henry Churchill, "What Kind of Cities Do We Want?" in Woodbury, ed., *The Future of Cities and Urban Redevelopment* (Chicago: University of Chicago Press), 50–51.

5. Douglass Lee, *Models and Techniques for Urban Planning* (Washington, D.C.: U.S. Department of Commerce National Bureau of Standards, 1968), 5–3; Arthur D. Little, *Community Renewal Programming: A San Francisco Case Study* (New York: F. A. Praeger, 1966).

6. Participants are listed in Garry Brewer, *Politicians, Bureaucrats, and the Consultant: A Critique of Urban Problem Solving* (New York: Basic Books, 1973), 112.

7. Steger went on to become president of the Urban and Regional Information Systems Association (URISA), which was created in the early 1960s, the professional association for the field. Although Lowry departed for RAND early in the CRP's model-building process, he maintained active links to Pittsburgh. An urban model he created at RAND was altered and used in the Pittsburgh CRP.

8. Charles Leven, *Potential Applications of Systems Analysis to Regional Economic Planning* (Pittsburgh: University of Pittsburgh Center for Regional and Economic Studies, 1964), VF NAC 7030.2 L.

9. David Brenneman, *Design for a Mathematical Model of Blight* (Pittsburgh: Pittsburgh Department of City Planning Community Renewal Program, 1962); Neiland Douglas and Wilbur Steger, *Choices in a Large Scale Modeling Effort: The Pittsburgh Simulation Model* (Pittsburgh: Community Renewal Program and CONSAD, 1964); Charles Leven and Bruce Newling, *Employment, Income, and Population Submodels* (Pittsburgh: University of Pittsburgh Center for Regional and Economic Studies, 1964); Steven Putnam, *Pittsburgh CRP, Industrial Location Model* (Pittsburgh: CONSAD, 1963); Pittsburgh Department of City Planning, *Data Processing and Simulation Techniques* (Pittsburgh: Department of City Planning Community Renewal Program, 1962).

10. Pittsburgh Department of City Planning (1962), n.p.

11. Pittsburgh Department of City Planning (1962).

12. Donald Gerwin, *A Proposed Model to Aid in Urban Renewal Choices* (Pittsburgh: Department of City Planning, December 1964).

13. Leven (1964).

14. Ira Lowry, *Seven Models of Urban Development* (Washington, D.C.: Institute for Applied Technology, September 1967), 45; Robert Yin, presentation in J. Bergsman and H. Wiener, eds., *Urban Problems and Public Policy Choices: Proceedings of the Symposium on Urban Growth and Development Sponsored by the Washington Operations Research Council and the Urban Institute, April 16–17, 1973* (New York: Praeger, 1975). An extensive critique of the Pittsburgh experiments by RAND affiliates is in Brewer (1973).

15. Lee (1968), 5–37; Douglass Lee, "Requiem for Large-scale Models," *JAIP* 39 (1973): 163–78.

16. Rioting did occur in Pittsburgh, but only after the 1962–64 period, when these efforts were under way.

17. Augur (May 5, 1946): 15–16. Operation Crossroads was the test of two nuclear weapons in summer 1946.

18. Augur (May 5, 1946); Augur, "Dispersal" (1948).

19. Barry Gottehrer, "Urban Conditions: New York City," in *Proceedings of the Urban and Regional Information Systems Association Annual conference* (Kent, Ohio: Center for Urban Regionalism, Kent State University, 1967), 2.

20. Kenneth Clark, "The Wonder Is There Have Been So Few Riots," *New York Times Magazine*, September 5, 1965, 10.

21. Translation of Hans-Herbert Gotz, "Mastering the World with Computers," from *Frankfurter Allgemeine Zeitung*, August 1, 1969, n.p., box NYC Rand Institute, folder *Rand New York Newsletter,* no. 6 (1969), RAND.

22. Kerner Commission [National Advisory Commission on Civil Disorders], *Report* (New York: Dutton, 1968).

23. Letter from Carmine Novick to heads of agencies, March 30, 1970, Mayor Lindsay Papers, subject files 1966–73, box 21, folder 365, Civilian Defense 1966–73, MANY.

24. Augur, "Dispersal" (1948), 35.

25. Yohuru Williams, "American Exported Black Nationalism: The Student Coordinating Committee, the Black Panther Party, and the Worldwide Freedom Struggle, 1967–1972," *Negro History Bulletin* 60, no. 3 (1997): 13–20; Mary Dudziak, *Cold War Civil Rights* (Princeton, N.J.: Princeton University Press, 2000).

26. Charles Haar, *Between the Idea and the Reality: A Study in the Origin, Fate, and Legacy of the Model Cities Program* (Boston: Little, Brown, 1975), 79.

27. Henry Maier, "The Troubled City," in Grace Kelleher, ed., *The Challenge to Systems Analysis: Public Policy and Social Change* (New York: Wiley, 1970), 57.

28. NYC Department of City Planning, *Application to the Housing and Home Finance Agency for a Community Renewal Program Grant* (New York: Department of City Planning, 1960), VF NAC 1613.5g27 NY.

29. NYC Mayor's Housing Executive Committee, *New York City's Renewal Strategy* (New York: Community Renewal Program, 1965), iv–v.

30. Wilbur Steger, "Development of a Management Information System for the New York City Planning Department," in *Proceedings of the Fourth Annual Conference on Urban Planning Information Systems and Programs* (Berkeley: University of California, 1966).

31. NYC Department of City Planning and CONSAD, "Development of a Management Information System for the New York City Planning Department," unpublished, September 7, 1965, cited in Steger (1966).

32. The executive director, unnamed, is quoted by Steger (1966), 24–25; see also M. C. Branch, "Comprehensive Planning: A New Field of Study," *JAIP* 25, no. 3 (1959): 115–20.

33. Steger (1966), 34.

34. NYC CRP, *Between Promise and Performance* (NYC: CRP, 1968).

35. Rowen turned down the job for a position at MIT and then was hired to become RAND president soon after: personal conversation with author, January 11, 2002.

36. Richard Rosenbloom and John Russell, *New Tools for Urban Management* (Boston: Harvard University Graduate School of Business Administration, 1971), and R. L. Petruschell, *Some Remarks on Planning, Programming, and Budgeting* (Santa Monica, Calif.: RAND, 1968). The latter compares budgeting in the U.S. Air Force and NYC's Health Services Administration.

37. Gottehrer (1967), 22; Timothy Cooney, *Reorganization of Civil Defense: A Report to John V. Lindsay* (New York: Office of Civil Defense, 1966).

38. E. S. Savas, R. Amsterdam, and E. Brodheim, "Creation of a Geographic Information System," in *Proceedings of the Association of Computing Machinery Conference on the Application of Computers to the Problems of Urban Society* (New York: ACM, 1969), 100.

39. Gottehrer (1967), 25.

40. Richard Reeves, "City Hires Rand Corp. to Study Four Agencies," *New York Times,* January 9, 1968, A1.

41. Reeves (1968), A1.

42. David Jardini, "Out of the Blue Yonder: The RAND Corporation's Diversification to Social Welfare Research," Ph.D. dissertation, Carnegie Mellon University, 1996; RAND Corp., *Report on a Study of Non-military Defense* (Santa Monica, Calif.: RAND, 1958).

43. Ira Lowry, *Report on PPBS Presentation to the New York City Housing and Development Administration* (Santa Monica, Calif.: RAND, February 26, 1968); Ira Lowry and Michael Teitz, *Programming Housing Improvement in New York City: Quantity and Quality* (Santa Monica, Calif.: RAND, 1969); Peter Rydell, *Landlord Reinvestment Model: A Computer Based Method of Evaluating the Financial Feasibility of Alternative Treatments for Problem Buildings* (Santa Monica, Calif.: RAND, 1970); David Shipler, "Rent Plan Close to RAND's Ideas," New York Times, May 17, 1970, sec. 8, 1.

44. Request for approval, official out-of-town trips form, Mayor Lindsay departmental files 1966–1973, box 8, folder 100, City Planning Commission 1966–67, MANY. That year, Elliott also traveled to Pittsburgh to inspect that city's work on redevelopment.

45. No box, folder NYC Rand Institute 1969, letter from Henry Rowen to John McCone, September 8, 1969, RAND.

46. John Kain, *A Proposed Rand Strategy for Domestic Policy Research: A Program of Research on the American Negro* (Santa Monica, Calif.: RAND, 1967); John Kain, *An Econometric Model of Metropolitan Development* (Santa Monica, Calif.: RAND, 1962); and John Kain, "A Computer Version of How a City Works," *Fortune,* November 1969, 242.

47. Memo, March 3, 1970, Joel Edelman to Peter Szanton, "A Broader Concept of the Mayor's Office Request for a 'Crisis Anticipation' System and Possible HUD Demographic Study Participation," no box, folder NYC Rand Institute 1970, RAND.

48. Memo, March 9, 1970, Robert Levine to Joel Edelman, "Mayor's Office Request for Assistance," no box, folder NYC Rand Institute 1970, RAND. This memo was widely

distributed. The March 9 memo was linked to an earlier discussion about developing early warning systems for the city.

49. Joan Jacoby, "DEWS—District Early Warning System for Neighborhood Deterioration," in *Proceedings of the Urban and Regional Information Systems Association Annual Conference* (Kent, Ohio: Center for Urban Regionalism, Kent State University, 1967): 38.

50. Ida Hoos, *Systems Analysis in Public Policy* (Berkeley: University of California Press, 1972), 109. A memo dated January 12, 1972, "Possible Sticky Tolchin Questions," included answers to questions about why the costs were so high for contracts with the city and why travel was so extensive for a city-based consulting group: box NYC Rand Institute, folder Memos/Notes/Statements, RAND.

51. Letter to Lindsay from Timothy Costello, deputy mayor-city administrator, February 3, 1969, Lindsay Papers, subject files 1966–1973, box 27, folder 477, Costello, Timothy W., Deputy Mayor–City Administrator (2), 1970–1972, 2, MANY.

52. Paul Dickson, "Think Tank Brings Systems Analysis to the City Streets," *Smithsonian* 5 (March 1975): 43–48.

53. Scott MacDonald, "Case History: Splintered City and a Fusion Mayor—New York's Lindsay," *Government Executive* 1 (June 1969); Frederick O'Reilly Hayes, "Managing the Unmanageable," in *Proceedings of the Urban and Regional Information Systems Association Annual Conference* (Akron, Ohio: URISA, 1970); "Savings Effected for the City of New York by the New York City–Rand Institute," folder Memos/Notes/Statements, RAND.

54. Summary of NYC–Rand Institute, "Work for the City of New York," May 30, 1972, 1, box NYC Rand Institute, folder Memos/Notes/Statements, RAND. This was the closest RAND came to any sort of city planning.

55. RAND, Twentieth Year Conferences, NYC special session, March 11, 1969, box NYC RAND Institute, folder Twentieth Year Conferences, 31.

56. The consultant controversies are described in a series of articles in the *New York Times*, June 7, 1970, 64; October 6, 1970, 55, 96; October 29, 1970, 1.

57. Leland Swanson and Glenn Johnson, eds., *The Cybernetic Approach to Urban Analysis* (Los Angeles: University of Southern California Graduate Program in City and Regional Planning, 1964), 27; T. Tamaru, "Prospects in Municipal Information Systems: The Example of Los Angeles," *Computers and Automation* 17 (January 1968): 15–18.

58. Harold Goldstein, *The Urban Information System: Some Concepts, Issues, and Experiences* (Cleveland, Ohio: Urban Studies Center, Batelle Memorial Institute, 1968), 40. Stanford Optner, *Report on the Feasibility of Electronic Data Processing in City Planning* (Los Angeles: S. L. Optner, 1959); D. S. Stroller and R. L. Van Horn, *Design of a Municipal Information System* (Santa Monica, Calif.: RAND, 1958).

59. System Development Corporation, *Development of an Automated Land Records File for Operations and Planning* (Santa Monica, Calif.: SDC, 1963); Advanced Information Systems, "General Proposal: Los Angeles Automated Plan File," January 25, 1963 (not published); and Computer Usage Company, "Proposal for the Mechanization of a City-wide Information Retrieval and Reporting System," April 11, 1963 (not published).

60. Calvin Hamilton, *Experience in Operation: Lecture at the Conference on Urban and Regional Information Systems for Public Policy Planning Operations in the United States* (Pittsburgh, Pa.: Pittsburgh Area Information Bank, August 20, 1965), VF NAC 830.

61. Swanson and Johnson (1964), 46.

62. Los Angeles Council file 77-2694, Municipal Archives of Los Angeles (hereafter, MALA), and Carter Bales, "The Program/Progress of Analysis and PPB in New York City Government," in Rosenbloom and Russell (1971).

63. Los Angeles Department of City Planning, *Mathematical Model Development Program: Introduction and Proposed Program* (Los Angeles: Department of City Planning, September 1966), 7.

64. *A Proposal for the Establishment of an Automated Planning and Operational File* (Los Angeles: Subcommittee on the Planning and Operational File, 1965), VF NAC 830g27 LosA L.

65. Ann Markusen, Peter Hall, Scott Campbell, and Sabrina Detrick, *The Rise of the Gunbelt* (New York: Oxford University Press, 1991).

66. Contracts 4251, 42518, 42519, 42520, MALA.

67. Harry Harmon, *Simulation: A Survey* (Santa Monica, Calif.: SDC, 1961); Herbert Isaacs, *Systems Considerations in Building a Metropolitan Data Bank for Urban Research* (Santa Monica, Calif.: SDC, 1962); Caleb Lanning and Herman Berkman, *Proceedings of the Conference on ADP in Local Governments* (Santa Monica, Calif.: SDC, 1965).

68. System Development Corporation, *Urban and Regional Information Systems: Support for Planning in Metropolitan Areas* (Washington, D.C.: GPO, 1968).

69. Ordinance no. 133,790, file no. 126368 (1967), MALA. Letters at MALA reveal that officials at the New York CRP and in Los Angeles at CAB were in touch; see, e.g., council file 71-968, MALA.

70. Jardini (1996), 413–14.

71. *Los Angeles Urban Information System Experience: A Brief Report* (Los Angeles: CAB, 1970).

72. *A Comprehensive and Efficient Method of Accessing Organizational Activity: The Community Program Information System* (Los Angeles: CAB, 1972).

73. *A Proposed Study Design for Computer Simulation to Support Studies of Urban Blight and Obsolescence* (Los Angeles: CAB, 1970), 13.

74. *Proposed Study Design for Computer Simulation* (1970), 2.

75. *LA Information System: Brief Report* (1970). 37. This CAB report differed from others in that it acknowledged that the thermostat metaphor was not a perfect fit, reminding readers that city temperature levels were frequently unknown or unquantifiable and that tools for measuring the performance of city government were not always precise.

76. *State of the City* (1970), 4:26.

77. Appendix to *Requirements and Constraints on the Design of the Urban Information System* (Los Angeles: CAB, 1970), 4.

78. Appendix to *Requirements and Constraints* (1970), 4.

79. Robert Joyce, "A City Looks at Technology Transfer," in *Second Annual Urban Technology Conference and Technical Display* (New York: American Institute of Aeronautics and Astronautics, 1972), 1.

80. *LA Information System: Brief Report* (1970), iii; *Proposed Study Design for Computer Simulation* (1970), 2.

81. *State of the City* (1970/71), 1:2; *State of the City* (1970/71), 4:8.

82. *Design Requirements for the Data and Systems Support Essential to an Urban Blight Systems Analysis: A Report* (1970), 9–10.

83. *LA Information System: Brief Report* (1970), 35.

84. Joyce (1972), 5.

85. Decision Sciences Corporation, *A Review of the State of the Art in Community Analysis, Simulation, and Modeling* (Jenkintown, Pa.: DSC, 1971).

86. Aaron Wildavsky, "The Political Economy of Efficiency: Cost Benefit Analysis, Systems Analysis, and Program Budgeting," *Public Administration Review* 26, no. 4 (1966): 292–310; Ida Hoos, *Systems Analysis in Public Policy* (Berkeley: University of California, 1972); Brewer (1973); and Alvin Drake, Ralph Kenney, and Philip Morse, eds., *Analysis of Public Systems* (Cambridge, MIT Press, 1972).

87. Yin (1975), 8.

88. Kenneth Clark and Jeannette Hopkins, *A Relevant War against Poverty* (New York: Harper, 1969), viii; James Kalish, "Flim-flam, Double Talk, and Hustle: The Urban Problems Industry," *Washington Monthly* 1 (November 1969): 6–16.

89. Abe Gottlieb, "The Computer and the Job Undone," in *Proceedings of the Association for Computing Machinery Conference on the Application of Computers to the Problems of Urban Society* (New York: ACM, 1970) 8–9.

90. Nicholas Meiszer, "The Effects of the USAC Project on Municipal Finance Function," is one of several papers at the 1970 URISA conference that discuss information systems as used in model cities: URISA, *Proceedings* (1970).

91. Gottlieb (1970), 6.

92. Swanson and Johnson (1964); Kenneth Kraemer and John Leslie King, *Computers, Power, and Urban Management* (Beverly Hills: SAGE, 1976); Roger Sission, in AAPSS (1967).

93. Leven (1964).

94. Anthony Downs, "PPBS and the Evolution of Planning," in *Proceedings of the Annual Conference of the American Society of Planning Officials* (Washington, D.C.: ASPO, 1967); Martin Kuenzlen, *Playing Urban Games* (Boston: I Press, 1972); Jack Dyckman, "The Scientific World of the City Planners," *American Behavioral Scientist* 6, no. 6 (1963): 46–51; and Robert Boguslaw, *The New Utopians* (Englewood Cliffs, N.J.: Prentice-Hall, 1965).

95. MacDonald (1969).

96. Swanson and Johnson (1964), 45.

97. John Kolesaar, "The States and Urban Planning and Development," in Alan Campbell, ed., *The States and the Urban Crisis* (New York: American Assembly, 1970), 135.

98. Downs (1967), 95.

99. Hoos (1972), 68.

100. Eugene Nickerson, "The Computer Backlash," in *Proceedings of the Urban and Regional Information Systems Association* (Kent, Ohio: Center for Urban Regionalism, Kent State University, 1967), 206.

101. William Ross, "Some Perspective on Federal PPBS," ASPO, *Proceedings* (1967); Kuenzlen (1972) and Jardini (1996) observed similar tensions brewing between systems experts and average citizens demanding a voice in the planning process for the Community Action Program.

102. Bernard Schriever, "Rebuilding Our Cities for People," *Air Force and Space Digest* (August 1968).

103. *Evaluation of the Urban Systems Engineering Demonstration Program, HUD Office of Community Planning and Development* (Washington, D.C.: GPO, 1974), III-2.

104. NYC special session, March 11, 1969, University Club, RAND Archives, box NYC RAND Institute, folder Twentieth Year Conferences, 23, RAND.

105. Yin (1975); Lee (1973).
106. Paul Starr, "Seductions of Sim: Policy as a Simulation Game," *American Prospect* 5, no. 17 (1994): 19–29.
107. Kenneth Kraemer and John Leslie King, "A Requiem for USAC," *Policy Analysis* 5, no. 3 (1979): 313–49.
108. For example, New York City contracted with McKinsey to survey the economic damage of the World Trade Center attacks.
109. Dickson (March 1975), 48.

FOUR: Urban Intelligence Gathering

1. Grace Kelleher, ed., *The Challenge to Systems Analysis: Public Policy and Social Change* (New York: Wiley, 1970), v.
2. Vice President Hubert Humphrey's speech at the Smithsonian, quoted in *Aerospace Technology* (May 20, 1968): 19. Following the Moon landing, the phrase "if we can put a man on the moon, we can————" became a popular expression of faith in American know-how.
3. Michael Sherry, *In the Shadow of War* (New Haven, Conn.: Yale University Press, 1995); James Hansen, "Demystifying the History of Aeronautics," in Martin Collins and Sylvia Fries, eds., *A Spacefaring Nation* (Washington, D.C.: Smithsonian, 1991).
4. William Burrows, *Deep Black: Space Espionage and National Security* (New York: Random House, 1986), 32.
5. Beaumont Newhall, *Airborne Camera: The World from the Air and Outer Space* (New York: Hastings House, 1969), 105.
6. M. C. Branch, *City Planning and Aerial Information* (Cambridge: Harvard University Press, 1971), 3–4. Many of these photographs can be located in the National Archives in record groups 373 and 263.
7. John Estes and Leslie Senger, eds., *Remote Sensing: Techniques for Environmental Analysis* (Santa Barbara, Calif.: Hamilton, 1974).
8. Kirk Stone, "Geography's Wartime Service," *Annals of the Association of American Geographers* (hereafter, *AAAG*) 69, no. 1 (1979): 89.
9. Norman Green, "Aerial Photographic Interpretation and the Social Structure of the City," *Photogrammetric Engineering* 23, no. 1 (1957): 90.
10. M. C. Branch, *Aerial Photography in Urban Planning and Research* (Cambridge: Harvard University Press, 1948), 8.
11. Norman Green and Robert Monier, *Reliability and Validity of Air Reconnaissance as a Collection Method for Urban Demographic and Sociological Information* (Maxwell Air Force Base, Alabama: Maxwell Air Force Base, 1953), iii. The turn toward aerial photography as a tool to document socioeconomic issues in a larger context was a more mathematical follow-on to a much older tradition—what was called social photography, sociological photography, or reform photography, dating to an earlier moment of urban crisis in the United States. In the 1880s, Jacob Riis, a journalist and social reformer who became well known for his studies of urban poverty in New York City, was perhaps the best-known user of photographic evidence alongside textual descriptions to bring the blight of tenement environments to his audiences. Riis pioneered the use of photography in the United States as an urban (and later rural) evidence-gathering technology, and other documentarians, including Lewis Hine, Dorothea Lange, and Walker Evans,

followed soon after, bringing this method to populations beyond NYC; see Maren Stange, *Symbols of Ideal Life: Social Documentary Photography in America, 1890–1950* (New York: Cambridge University Press, 1989).

12. Pamela Mack, *Viewing the Earth* (Cambridge: MIT Press, 1990), 32.

13. Merton Davies and William Harris, *RAND's Role in the Evolution of Balloon and Satellite Observation Systems and Related US Space Technology* (Santa Monica, Calif.: RAND, 1988).

14. John Cloud, "Imaging the World in a Barrel: CORONA and the Clandestine Convergence of the Earth Sciences," *Social Studies of Science* 31, no. 2 (2001): 231–51; Mary Rabbitt, *A Brief History of the U.S. Geological Survey* (Reston, Va.: USGS, 1984); John Cloud, "Crossing the Olentangy River," *Studies in the History and Philosophy of Modern Physics* 31, no. 3 (2000):371–404.

15. Charles Andregg, "Military-USGS Mapping Cooperation," *Photogrammetric Engineering and Remote Sensing* 45, no. 12 (December 1979): 1623.

16. Arthur Robinson, "Geography and Cartography Then and Now," *AAAG* 69, no. 1 (1979): 97. At the National Archives, record group 57, "Aerial Photography of the Geological Survey, 1935 to 1942," is cross-linked to record group 373, "Aerial Photography of the Defense Intelligence Agency."

17. Howard McCurdy, *Inside NASA: High Technology and Organizational Change in the U.S. Space Program* (Baltimore, Md.: Johns Hopkins University Press, 1993); Adam Yarmolinksy, *The Military Establishment: Its Impacts on American Society* (New York: Harper & Row, 1971), 11.

18. Mack (1990), chapter 4; McCurdy (1993), 51.

19. Johnson, cited in Burrows (1986), vii.

20. Leonard Bowden and Darryl Goehring, *Remote Sensing in County Planning: A Case History from Los Angeles County* (Washington, D.C.: USGS, 1970), 10.

21. Robert Hotz, "The Turbulent Summer," *Aviation Week and Space Technology* (July 31, 1967): 11.

22. James Webb, *Space Age Management: The Large-scale Approach* (New York: McGraw-Hill, 1969).

23. This idea was also supported by Webb's predecessor, T. Keith Glennan, in his report *The Usefulness of Aerospace Management Techniques in Other Sectors of the Economy* (Santa Monica, Calif.: RAND, 1964).

24. Webb (1969), 67.

25. Carl Feiss, "The Foundations of Federal Planning Assistance," *Journal of the American Planning Association* 51, no. 2 (1985): 174–84.

26. National Committee on Urban Growth Policy, *The New City* (New York: Praeger, 1969); U.S. Domestic Council, *The President's 1971 Environmental Program: Reform Renewal for the 70's* (Washington, D.C.: GPO, 1971).

27. Douglas Yates, *The Ungovernable City* (Cambridge: MIT Press, 1977); Harold Rothbart, "The Utilization of Technological Innovations in Urban Management," *Third Annual Urban Technology Conference and Technical Display* (New York: AIAA, 1973).

28. Webb (1969), 10–11.

29. NASA, *Conference on Space, Science, and Urban Life* (Washington, D.C.: GPO, 1963).

30. NASA (1963), 1.

31. Webb had long expressed his commitment to finding ways to apply talented students working in science to problems of national significance. Earlier in his career, be-

fore joining NASA, Webb created a Frontiers of Science Foundation in his home state of Oklahoma to increase the number of students studying science in the service of national priorities.

32. Dickson (1971), 215; Bernard Schriever, "Rebuilding Our Cities for People," *Air Force and Space Digest* (August 1968).

33. Thomas Paine, "The City as an Information Network," *IEEE International Convention Record* (New York: IEEE, March 1966).

34. Amrom Katz, "A Retrospective on Earth Resource Surveys: Arguments about Technology, Analysis, Politics, and Bureaucracy," *Photogrammetric Engineering* 42, no. 2 (1976): 197.

35. S. R. Siegel, *Management Technology Applied to Urban Systems, Proceedings of the Symposium on Application of NASA Management Technology to Management of Urban Systems, April 12–13, 1972* (Philadelphia: Drexel University Center for Urban Research and Environmental Studies, 1972), 43.

36. Simon Ramo, *Cure for Chaos: Fresh Solutions to Social Problems through the Systems Approach* (New York: D. McKay, 1969).

37. L. F. Eastwood, J. K. Gohagan, C. T. Hill, R. P. Morgan, S. M. Bay, T. K. Foutch, T. R. Hays, *Final Report—Appendices, Project on Earth Observation Data Management Systems* (St. Louis: Washington University, December 31, 1976), 44.

38. Robert Alexander, "Multispectral Sensing of Urban Environments," in *Five Papers on Remote Sensing and Urban Information Systems, Technical Report 1* (Evanston, Ill.: Northwestern University Department of Geography, April 1966), n.p.

39. Branch (1971), 3.

40. Eric Moore and Barry Wellar, "Urban Data Collection by Airborne Sensor," *JAIP* 35 (1969): 35–43; D. E. Anderson and P. N. Anderson, "Population Estimates by Humans and Machines," *Photogrammetric Engineering* 39, no. 2 (1973): 147–54; Frederick Doyle, "Can Satellite Photography Contribute to Topographic Mapping?" in Robert Hotz, *The Surveillant Science: Remote Sensing of the Environment* (Boston: Houghton Mifflin, 1973), 153; Craig Tom, Lee Miller, and Jerrold Christenson, *Spatial Land-use Inventory, Modeling, and Projection, Denver Metropolitan Area, with Inputs from Existing Maps, Airphotos, and Landsat Imagery* (Maryland: NASA Goddard Space Flight Center, August 1978), 5; David Nichols and William Brooner, *Interfacing Remote Sensing and Automated Geographic Information Systems* (Riverside: University of California Department of Geography, 1972); David Lindgren, "Urban Applications of Remote Sensing," in Estes and Senger (1974): 225.

41. Remarks at House Appropriations Committee, Hearings, agricultural subcommittee hearing on fiscal 1968 budget, 242, cited in Edmund Lambeth, "Eyes in Orbit: The Potential, Problems, and Prospects of Monitoring Earth from Space," February 1968, in papers by participants in the science, technology, and public policy seminar, 1968–72, Harvard University Kennedy School of Government library collections, unpublished materials.

42. Frank Victor Westerlund, "Modes of Application of Remote Sensing Technology and Media to Urban and Regional Planning, with Emphasis on Surface Classification and Alternative, Problem-Oriented Approaches," Ph.D. dissertation, University of Washington, 1977; Harry Mallon, Joan Howard, and Kenneth Karch, *An Examination of Applications of Remote Sensing Data to Metropolitan Washington Council of Governments Planning Requirements* (Washington, D.C.: USGS, 1971); Harry Mallon and Joan Howard,

An Assessment of Remote Sensor Imagery in the Determination of Housing Quality Data. (Washington, D.C.: USGS, 1971).

43. Duane Marble and Frank Horton, "Extraction of Urban Data from High and Low Resolution Images," in *Proceedings of the Sixth International Symposium on Remote Sensing of Environment* (Ann Arbor: University of Michigan, 1969), vol. 2, 814; Robert Colwell, Paul Lowman, and Sandra Wenderoth, *Apollo 9 Multispectral Photographic Information* (Washington, D.C.: NASA, April 1970); Robert Simpson, "Radar: Geographic Tool," *AAAG* 56 (1966): 80–96; Robert Rudd and Richard Highsmith, *The Use of Air Photo Mosaics as Simulators of Spacecraft Photography in Land Use Mapping* (Washington, D.C.: USGS, 1970).

44. H. G. Goodell and W. Reed, *The Potential of Remote Sensing as a Data Base for State Agencies: The Virginia Model* (Charlottesville: University of Virginia Department of Environmental Sciences, 1971); Derek Thompson, "Small Area Population Estimation Using Land Use Data Derived from High Altitude Aircraft Photography," in *Proceedings of the American Society of Photogrammetry* (Falls Church, Va.: American Society of Photogrammetry, 1975), 673–96; Joseph Sabol, "The Relationship between Population and Radar Derived Area of Urban Places," in *The Utility of Radar and Other Remote Sensors in Thematic Land Use Mapping from Spacecraft* (Washington, D.C.: USGS, 1968), 46–74; K. E. Foster and J. D. Johnson, *Research for Applications of Remote Sensing to State and Local Governments*, annual report (Tucson: University of Arizona Office of Arid Land Studies, February 1973); Ellen Knapp and Deborah Rider, "Automated Geographic Information Systems and Landsat Data: A Survey," in Harvard University Laboratory of Computer Graphics and Spatial Analysis, *Computer Mapping in Natural Resources and the Environment, Including Applications of Satellite Derived Data* (Cambridge: Harvard University Laboratory of Computer Graphics and Spatial Analysis, 1979), 59.

45. Bromberg singles out 1969 as a particularly interesting year for NASA, a year of contradiction, when the man-on-the-Moon landing offered a public face of success and yet Richard Nixon's arrival as president led to major budget cuts at the agency: Joan Lisa Bromberg, *NASA and the Space Industry* (Baltimore: Johns Hopkins University Press, 1999).

46. *Aerospace Technology* (May 20, 1968): 19.

47. Cited in Hoos (1972), 87. The full statement by Nelson appears in *Congressional Record* (October 18, 1965).

48. Torrey (1967), 31–32. Two later summer studies, in 1967 and 1968, focused on the benefits of Earth satellites.

49. Torrey (1967), 37–38.

50. "NASA Aide Named to Housing Post," *New York Times,* March 16, 1969. A few years earlier, HUD appointed Dwight Ink to be assistant secretary of HUD. Ink moved to HUD from the Atomic Energy Commission, where from 1959 to 1966 he had served as assistant general manager. "Johnson Shifts A.E.C. Official," *New York Times,* February 16, 1966, 46.

51. Siegel (1972), 38–39.

52. *Aerospace Orientation Project* (Berkeley, Calif.: Aerospace Orientation Program, 1971); University of California, Berkeley College of Environmental Design, *Aerospace Employment Project* (Berkeley, Calif.: Aerospace Orientation Program, 1972); *Project ADAPT: Report #1* (Cambridge: MIT, Department of Urban Studies and Planning, 1972); Francis Ventre and Larry Sullivan, *Project ADAPT: Report #2* (final report) (Cambridge: MIT, Department of Urban Studies and Planning, May 1972).

53. Specific placements of workers following the program are detailed in the MIT May 1972 report.

54. *UTS Briefs;* Public Technology, Inc., *The Urban Technology System: A Positive Strategy for Technology Transfer* (Washington, D.C.: PTI, 1977); P. Buchman, "Application of Satellite Remote Sensing to Local Governments through Urban Technology System Network," *Photogrammetric Engineering and Remote Sensing* 44, no. 6 (1978), 742.

FIVE: Moon-Shot Management for American Cities

1. William Mitchel, "The Cities Can't Pay for Aerospace Technology," in Alan Westin, ed., *Information Technology in a Democracy* (Cambridge: Harvard University Press, 1971), 455.

2. George Goddard, *Overview: A Life-long Adventure in Aerial Photography* (Garden City, N.Y.: Doubleday, 1969).

3. Willis Lee, *The Face of the Earth as Seen from the Air* (New York: American Geographical Society, 1922); Guy Hayler, "The Aeroplane and City Planning: The Advantages of Viewing Cities from Above," *American City* 23, no. 6 (1920): 575–79; and Nelson Lewis, "A New Aid in City Planning," *American City* 26, no. 3 (1922): 209–12.

4. Fairchild Aerial Surveys, "Satisfy Your Taxpayers with Completely Detailed Maps," advertisement, 1928, VF NAB 1244 MICRO F101008; Sherman Fairchild, "The Aerial Survey in Connecticut Tax Equalization," *Connecticut Industry*, April 1928, 5–9; Sherman Fairchild, "Engineering from the Air," advertisement, December 16, 1927, VF NAB 1244; Fairchild Aerial Surveys, "Aerial Photography in War and Peace," advertisement, ca. 1920s, VF MICRO; Aero Service Corporation, *Advanced Techniques of Data Collection for City and Regional Planning* (Philadelphia: Aero Service, 1961).

5. Charles Emerson, "Aerial Surveying as Applied to Tax Equalization," *American City* 36, no. 4 (1927): 522–24; Mississippi State Planning Commission, *Special Report on Aerial Survey to Tax Assessor Association of Mississippi* (Jackson: [Planning Commission], 1936), VF NAC 839g 25 Mis.

6. "Solving Other Municipal Problems with Fairchild Aerial Surveys," advertisement, *Connecticut Industry* (April 1928); 4; Mississippi State Planning Commission (1936).

7. Tennessee Valley Authority, *The Rural Land Classification Program* (Washington, D.C.: TVA, 1935); F. J. Marschner, *Land Use and Its Patterns in the United States* (Washington, D.C.: GPO, 1959); USGS Map Information Office, *Status of Aerial Photography in the United States* (December 31, 1946), VF NAC 1244 U; N. Carls, *How to Read Aerial Photographs for Census Work* (Washington, D.C.: GPO, 1947).

8. Matthew Witenstein, "Photosociometrics—the Application of Aerial Photography to Urban Administration and Planning Problems," *Photogrammetric Engineering* 20 no. 4 (1954): 419–27; G. A. Stokes, "The Aerial Photograph: A Key to the Cultural Landscape," *Journal of Geography* 49 (1950): 32–40; L. L. Pownall, "Aerial Photographic Interpretation for Urban Land Use in Madison, Wisconsin," *Photogrammetric Engineering* 26 (1950): 414–26; Ernest Metivier and Roger McCoy, "Mapping Urban Poverty Housing from Aerial Photographs," in *Proceedings of the Seventh International Symposium on Remote Sensing of the Environment* (Ann Arbor: University of Michigan, 1971), vol. 2., 1563–96; Rolt Hammond, *Air Survey in Economic Development* (New York: Elsevier, 1967); Leo Silberman, "Air Social Surveys: Principles and Developments," *Rural Sociology* 27:4

(1962): 475–83; Shane Davies, Alex Tuyahov, and Robert Holz, "Use of Remote Sensing to Determine Urban Poverty Neighborhoods," in Robert Holz, *The Surveillant Science: Remote Sensing of the Environment* (Boston: Houghton Mifflin, 1973), 386–90.

9. Photographs from the Regional Planning Federation of the Philadelphia Tri-State District (1928–52) can be found at the Temple University Urban Archives; Boston Finance Commission, "Letter to the Mayor Describing Maps for Use in City Planning" (Boston: [Boston Finance Commission], 1965), VF Ref Bos; Marjorie Rush and Sally Vernon, "Remote Sensing and Urban Public Health," *Photogrammetric Engineering and Remote Sensing* 41, no. 9 (1975): 1149–55; School of Public Health at University of Texas at Houston, *Identifying Urban Environmental Health Areas by Environmental Data Gathered by Remote Sensing* (Houston: University of Texas, 1972).

10. New Hampshire State Planning and Development Commission, *A Plan and Policy for Aerial Photography in New Hampshire* (Concord: State Planning and Development Commission, February 1949), 1, VF VAC 839g25 NH; Pennsylvania Department of Commerce, *Aerial Photographs as Tax Maps* (Harrisburg: Pennsylvania Department of Commerce, 1952).

11. New Hampshire State Planning and Development Commission (February 1949), 1; Abrams Aerial Survey Corporation, *Aerial Surveys and Maps from Photographs* (1952), 1, VF NAC 1224 A; Jeannette Davis, *Uses of Air Photos for Rural and Urban Planning, Department of Agriculture Handbook n315* (Washington, D.C.: GPO, 1966), 3.

12. Sherman Fairchild, "Winged Surveyors: What Aerial Photography Is Doing for Industry and Science," *Scientific American* (March 1922): 159; Fairchild Aerial Camera Corporation, *Aerial Surveying and Its Relation to City Planning* (1923), VF NAC 839.

13. Fairchild Aerial Camera Corp. (1923), 2; Lewis (March 1922); H. I. Brock, "The City that the Air Traveler Sees," *New York Times Magazine,* March 11, 1928: 14–15, 27; Fairchild (March 1922): 219.

14. Airmap Corporation of America, *Aerial Views of New York City, 1927–1939* (Brooklyn: Airmap, 1940); William Fried, *New York in Aerial Views* (New York: Dover, 1980); Bart Barlow, "Seeing Ourselves as the Airplanes See Us," *New York Times,* February 6, 1977, sec. 21, 21.

15. Ashraf Manji, *The Uses of Conventional Aerial Photography in Urban Areas* (Washington, D.C.: USGS, 1968), 1; "The Conservation of Existing Facilities for the Relief of Traffic Congestion," *Municipal Engineer's Journal* 8, no. 3 (1922): 132; "Traffic Arteries In and Out of Manhattan—the City's Battle with a Great Problem," *New York Sun* (January 14, 1929), n.p., VF NAC 6827 NY; East Midtown District, *Preliminary Report on Planning* (New York: East Midtown District, January 1939), VF NAC 6827 NY; NYC Planning Commission, *Proposed First and Second Stages of the Master Plan of Land Use* (New York: City Planning Commission, 1940), VF NAC 6827 NY; NYC Department of Parks, *The Future of Jamaica Bay* (New York: Moore Press, 1938; Robert Moses, "The Changing City," *Architectural Forum* (March 1940): 142–56; Robert Moses, *NYC Waterfront* (New York: Comet Press, 1948), VF NAC 6827 NY; *Openings . . . Federal, State, City Arterial Construction Program,* New York, 1950, VF NAC 6827 NY; NYC Committee on Slum Clearance, *Title 1 Slum Clearance Progress* (New York: [Committee on Slum Clearance] July 15, 1957), VF NAC 1433.8g27 New Y 1957; NYC Mayor's Housing Executive Committee (1965); Community Renewal Program (December 1968); letter from Donald Elliott, chair, NYC Planning Commission, to Howard Leary, police commissioner, March 13,

1967, Lindsay Papers, department files 1966–73, box 8, folder 101, City Planning Commission, 1967, MANY.

16. Keith Clarke and John Cloud, "On the Origins of Analytical Cartography," *Cartography and Geographic Information Science* 27, no. 3 (2000): 198.

17. Emanuel Savas, first deputy city administrator under Mayor Lindsay (1967–72), "Background Material—Management Science," Lindsay Papers, subject files 1966–1973, box 99, folder 1853, Dr. Emanuel Savas—First Deputy City Administrator. MANY.

18. E. S. Savas, R. Amsterdam, and E. Brodheim, "Creation of a Geographic Information System," in *Proceedings of the Association for Computing Machinery Conference on the Application of Computers to the Problems of Urban Society* (New York: ACM, 1969), 100–101; Robert Amsterdam and E. Brodheim, "Creation of a Geographic Information System," in *Proceedings of the Association of Computing Machinery Conference on the Application of Computers to The Problems of Urban Society* (New York: ACM, 1970); Robert Amsterdam, *GIST: A Geographic Information System for New York City: Preliminary Design* (New York: Office of the Mayor, 1968).

19. Request for approval, official out-of-town trips form, Lindsay Papers, departmental files 1966–1973, box 8, folder 100, City Planning Commission, 1966–67, MANY.

20. Letter to John Lindsay from Timothy Costello, January 5, 1970, Lindsay Papers, subject files 1966–73, box 27, folder 477, Costello, Timothy W., Deputy Mayor-City Administrator (2) 1970–72, MANY; Amsterdam (May 1971); Robert Amsterdam, "Guidelines for New York City's Community Shelter Plan," in *Proceedings of the Urban and Regional Information Systems Association Conference* (Kent, Ohio: Kent State University Center for Urban Regionalism, 1970); Elihu Levine, "A Computerized Community Shelter Plan," in *Proceedings of the Association for Computing Machinery Conference on the Application of Computers to the Problems of Urban Society* (New York: ACM, 1969); E. Brodheim and C. Wang, "Application of a Geographic Information System to Computerized Districting," in *Proceedings, Conference on Application of Computers* (1969); Robert Amsterdam, *An Introduction to GIST* (New York: Office of the Mayor, May 1971).

21. To Mayor Lindsay from Timothy Costello, deputy mayor–city administrator, re. NASA, May 27, 1968, and attached list of attendees, Lindsay Papers, subject files 1966–1973, box 27, folder 476, Costello, Timothy W., Deputy Mayor-City Administrator (1), 1966–69, MANY.

22. Letter to Jay Kriegel from Eward Skloot, assistant to the mayor, "Application of Technology to Urban Problems," December 2, 1971, 5, MANY; Lindsay Papers, subject files 1966–73, box 23, folder 395, Committee on Science and Technology 1972, MANY.

23. "Draft of Announcement by Mayor," Lindsay Papers, subject files 1966–73, box 23, folder 395, Committee on Science and Technology 1972, MANY.

24. "Draft of Letter from Mayor to Agency Heads," Lindsay Papers, subject files 1966–73, box 23, folder 395, Committee on Science and Technology 1972, MANY.

25. A. Karen, D. Orrick, and T. Anuskiewicz, "Technology Transfer in New York City: The NASA/NYC Applications Project," *Third Annual Urban Technology Conference* (New York: American Institute of Aeronautics and Astronautics, 1973), 1; Lindsay Papers, subject files 1966–73, box 23, folder 395, Committee on Science and Technology 1972, MANY.

26. Lindsay Papers, subject files 1966–73, box 23, folder 395, Committee on Science and Technology 1972, MANY.

27. Emanuel Savas, letter of resignation to Lindsay, October 16, 1972, Lindsay Papers, subject files 1966–73, box 99, folder 1853, Dr. Emanuel Savas—First Deputy City Administrator, MANY.

28. Arthur Row, *Reconnaissance of the Tri-State Region* (New York: Tri-State Transportation Commission, 1965); System Development Corp. (1968), 91–100; Tri-State Regional Planning Commission *Regional Development Guide, 1977–2000* (New York: Tri-State Regional Planning Commission, 1978).

29. Scott Bollens, "State Growth Management: Intergovernmental Frameworks and Policy Objectives," *Journal of the American Planning Association* 58 (1992): 454–66; Richard Barrows, *The Roles of Federal, State, and Local Governments in Land-use Planning* (Washington, D.C.: National Planning Conference, 1982); Fred Bosselman and David Callies, *The Quiet Revolution in Land Use Control* (Washington, D.C.: GPO, 1971).

30. T. P. Ahrens, *The Utilization of Aerial Photographs in Mapping and Studying Land Features* (Washington, D.C.: Resettlement Administration, October 1936).

31. Cornell University Center for Aerial Photographic Studies, *New York State Land Use and Natural Resources Inventory* (Ithaca, N.Y.: Cornell University Center for Aerial Photographic Studies, May 1968); New York State Office of Planning Services, *Land Use and Natural Resources Inventory of New York State, LUNR Classification Manual* (Albany: New York Office of Planning Services, June 1972).

32. Carl Steinitz et al., *New York State Natural Resources Inventory* (Albany: New York State Office of Planning Coordination, 1968); Harvard University Department of Landscape Architecture Research Office, *A Comparative Evaluation of Resource Analysis Methods* (Cambridge: Harvard University Department of Landscape Architecture, 1969).

33. Ronald Shelton and Ernest Hardy, "The New York State Land Use and Natural Resources Inventory," in *Proceedings of the Seventh International Symposium on Remote Sensing of Environment* (Ann Arbor: University of Michigan, 1971), vol. 2, 1571–75.

34. William Horne, "What Is LUNR?" in *Proceedings of the Kodak Seminar: Aerial Photography as a Planning Tool* (Rochester, N.Y.: Eastman Kodak, 1974).

35. Ernest Hardy, *Enhancement and Evaluation of Skylab Photography for Potential Land Use Inventories* (Ithaca, N.Y.: Cornell University Department of Natural Resources, 1975); Shelton and Hardy (1971), 1572; Ernest Hardy, "The Design, Implementation, and Use of a Statewide Land-use Inventory: The New York Experience," in *Proceedings of the NASA Earth Resources Survey Symposium, Houston, Texas* (Houston: NASA, Lyndon Johnson Space Center, 1975), vol. 1-C, 1573–77.

36. F. M. Huddleston, *Album of the Aerographic Company* (Los Angeles: Aerographic, 1930); Los Angeles Department of City Planning, *Bird's Eye View of Los Angeles* (Los Angeles: Department of City Planning, 1941); Judith Norvell Jamison, *Coordinated Public Planning in the Los Angeles Region* (Los Angeles: UCLA Bureau of Governmental Research, June 1948), 82; Manji (1968), *A View of Los Angeles Traffic* (Los Angeles: Foster & Kleiser, 1965); "Century City: An Alcoa Development," September 1, 1963, VF NAC 1613g27 Los A; Bowden and Goehring (1970). Contracts with individual city agencies, including 104810, 90191, 114656, 136004, 132833, 131953, and c-10937, can be found in the contracts files at MALA.

37. M. C. Branch, *City Planning and Aerial Information* (Cambridge: Harvard University Press, 1971).

38. *Applying Technology to Unmet Needs* (Washington, D.C.: GPO, 1966).

39. George Ervin and Lloyd Blomeyer, *First Annual Report . . . (Four Cities Program), Interagency Agreement Between NSF and NASA, July 71–Oct 72* (Pasadena, Calif.: JPL, 1972); Hoos (1972), 3; Edmund Brown, "Aerospace Studies for the Problems of Men," *State Government* 39 (winter 1966): 2.

40. Robert Joyce, "A City Looks at Technology Transfer," in *Second Annual Urban Technology Conference and Technical Display* (New York: American Institute of Aeronautics and Astronautics, 1972), 1.

41. *Los Angeles Housing Model, Technical Report* (Los Angeles: CAB, 1974), 161–62.

42. Robert Mullens, "Color Infrared Photography—a New Means of Detecting Urban Blight: Urban Technology Conference," in *Second Urban Technology Conference* (1972), 1.

43. Norman Thrower, Robert Mullens, and Leslie Senger, "Analysis of NASA Color Infrared Photography of Los Angeles," in Richard Pascucci and Gary North, eds., *Mission 73 Summary and Data Catalog* (Houston: Manned Spacecraft Center, 1968), 259–66; Robert Mullens and Leslie Senger, *Analysis of Urban Residential Environments Using Color Infrared Aerial Photography* (Riverside: University of California Department of Geography, 1969).

44. Charles Johnson and Robert Mullens, *A Practical Method for the Collection and Analysis of Housing and Urban Environment Data: An Application of Color Infrared Photography* (Los Angeles: CAB, 1970), v.

45. Johnson and Mullens (April 1970), 79.

46. Johnson and Mullens (April 1970), 29.

47. *LA Housing Model, Tech. Report* (1974), 161–62.

48. *The State of the Tenth Council District* (Los Angeles: CAB, 1971), 32.

49. Mullens (1972), 3.

50. *The Los Angeles Urban Information System Experience: A Brief Report* (1970), 23.

51. Joyce (1972), 6.

52. *LA Housing Model, Tech. Report* (1974), 1.

53. *LA Housing Model, Tech. Report* (1974), 159.

54. Robert Joyce, "A Practical Method for the Collection and Analysis of Housing and Urban Environment Data: Application of Color Infrared Photography," in *Proceedings of Kodak Seminar* (1974), n.p.

55. *LA Model, Tech. Report* (1974), 158; *State of the City*, vol. 2 (Los Angeles: CAB, 1975); *Transition Areas, Preliminary*, "not for public release" (Los Angeles: CAB, 1976), MALA.

56. *Los Angeles Housing Model, Summary Report* (Los Angeles: CAB, 1974), iv; *LA Housing Model, Tech. Report* (1974), 100; *Proceedings of Kodak Seminar* (1974). Robert Joyce, whose paper was presented by Robert Mullens after Joyce's retirement from CAB, was the only city manager represented.

57. Johnson and Mullens (April 1970), 87; Mullens (1972), 2; Albert Landini, "Policy Implications in Developing a Land Use Management Information System," and Charles Paul, "A Land Use Management Information System Implemented in Los Angeles," both in *Proceedings of the NASA Earth Resources Survey Symposium* (Houston: NASA, 1975).

58. Nevin Bryant, Albert Zobrist, and Albert Landini, memo, n.p., dated November 1975, following appendix A, in *LUMIS, Land Use Management and Information Systems* (Pasadena: JPL, 1976); Charles Paul, Albert Landini, and C. Deigert, "Remote Sensing Applications for Urban Planning, LUMIS Project," *Photogrammetric Engineering and Remote Sensing* 41, no. 6 (1975): 781 ff.

59. Charles Paul and Albert Landini, *Los Angeles City and Jet Propulsion Laboratory Documentation: Final Report, LUMIS—Land Use Management Information System* (Pasadena, Calif.: Jet Propulsion Laboratory, June 1975); Bryant, Zobrist, and Landini memo, November 1975.

60. Paul and Landini (June 1975), 13, table 2; Los Angeles Department of City Planning, *An Introduction to the Los Angeles Land Use Planning and Management Subsystem (LUPAMS)* (Los Angeles: Department of City Planning, January 1975); Los Angeles Department of City Planning, *Citywide Parcel Information System, Systems Design and Operations Manual* (Los Angeles: Department of City Planning, October 1974).

61. Letter from Mayor Bradley to city council, March 11, 1975, council file 75-1153, MALA. All city department heads were to attend four-day seminars in December 1974 and January 1975 so that they would be oriented to the city's new direction in municipal information system development.

62. William Donaldson, "A User's View of Technology Transfer," in *Aerospace Technology Transfer to the Public Sector* (New York: AIAA, 1978), 17–18.

63. William Ross, "Some Perspectives on Federal PPBS," in *Proceedings of the Annual Conference of the American Society of Planning Officials* (Washington, D.C.: ASPO, 1967), 89; Joyce (1972), 3–4.

64. Hoos (1972), 25.

65. Joyce (1972), 1–2; Ventre and Sullivan (May 1972), 3–4.

66. Urban Institute, *The Struggle to Bring Technology to Cities* (Washington, D.C.: Urban Institute, 1971), 11.

67. James Webb, *Space Age Management: The Large-scale Approach* (New York: McGraw-Hill, 1969), 59.

68. Finger's contributions to the project did receive public recognition: he became a fellow of the American Institute of Architects in 1977.

69. *Remote Sensing Applied to Urban and Regional Planning: Citations from the NTIS Database* (Washington, D.C.: NTIS, 1984).

70. William Miller, "Industry's Growing Space Harvest," *Industry Week* 197, no. 3 (May 1, 1978): 45–49.

71. *Proceedings of the Kodak Seminar* (1974).

72. Abrams Aerial Survey Corp. (1937); Davis (1966).

73. Two works have offered competing interpretations of how NASA made its choice: see Pamela Mack, *Viewing the Earth* (Cambridge: MIT Press, 1990), and John Cloud, "Re-viewing the Earth: Remote Sensing and Cold War Clandestine Knowledge Production," *Quest: The History of Space Flight Quarterly* 8, no. 3 (2000): 5–15.

74. *NASA Technical Memorandum TM X-58121, The ERTS-1 Investigation*, vol. 5: *Urban Land Use Analysis (Report for Period July 1972 to June 1973)* (Houston: Lyndon Johnson Space Center, October 1974).

75. Herbert Muschamp, "Assembling a Montage of Images in Glass and Steel," *New York Times*, January 27, 2002, 38.

76. By contrast, despite the privatization of Landsat and improvements to image resolution, recent analysts have observed it continues to be of limited use for planning applications in urban areas. In *Remote Sensing and Urban Analysis* (New York: Taylor & Francis, 2001) Jean-Paul Donnay, Michael Barnsley, and Paul Longley report a continuing disconnect between scientific research and planners' everyday needs.

77. Webb (1969), 104–5.

264   Notes to Pages 159–168

78. Mason Caldwell, cited in Siegel (1972), 98.

79. Julius Lester in Stewart Benedict, ed., *Backlash: Black Protest in Our Time* (New York: Popular Library, 1970), 292.

SIX: Cable as a Cold War Technology

1. Paul Baran, *On Distributed Communications* (Santa Monica, Calif., RAND, 1964).

2. Paul Baran and Martin Greenberger, *Urban Node in the Information Network* (Santa Monica, Calif.: RAND, 1967), 10. Baran is listed among the ten participants from RAND at a summer program on transportation that brought together many experts in housing and urban studies, including several HUD officials, as part of planning for RAND's new urban research program: cited in David Jardini, "Out of the Blue Yonder: The RAND Corporation's Diversification into Social Welfare Research, 1946–1968," Ph.D. dissertation, Carnegie Mellon University, 1996, 392 n. 74.

3. Leila Rupp, *Mobilizing Women for War: German and American Propaganda, 1939–1945* (Princeton, N.J.: Princeton University Press, 1978); Christopher Simpson, *Science of Coercion* (New York: Oxford University Press, 1994).

4. Robert McNamara, Montreal speech of May 18, 1966, cited in Robert McNamara, *The Essence of Security* (New York: Harper & Row, 1968), 142; Christopher Simpson, *Universities and Empire: Money and Politics in the Social Sciences during the Cold War* (New York: Free Press, 1998); Christopher Lasch, *The Agony of the American Left* (New York: Vintage, 1969); Michael Latham, *Modernization as Ideology: American Social Science and "Nation Building" in the Kennedy Era* (Chapel Hill: University of North Carolina Press, 2000); Seymour Deitchman, *The Best Laid Schemes: A Tale of Social Research and Bureaucracy* (Cambridge: MIT Press, 1976); Michael Klare, *War without End* (New York: Vintage, 1972); and Adam Yarmolinksy, *The Military Establishment: Its Impacts on American Society* (New York: Harper & Row, 1971).

5. Jardini (1996), 288.

6. Irene Gendzier, "Play It Again Sam: The Practice and Apology of Development," in Simpson (1998); Simpson (1994); Latham (2000).

7. Daniel Lerner and Wilbur Schramm, *Communication and Change in the Developing Countries* (Honolulu: East-West Center Press, 1967), Lucian Pye, ed., *Communications and Political Development* (Princeton, N.J.: Princeton University Press, 1963); and Harold Lasswell, Daniel Lerner, and Hans Speier, *Propaganda and Communication in World History*, vol. 2 (Honolulu: East-West Center, 1980).

8. Erwin R. Brigham, *Pacification Measurement in Vietnam: The Hamlet Evaluation System*, prepared for the SEATO Internal Security Seminar, Manila, June 3–10, 1968 (Vietnam: CORDS, 1968), 27.

9. Amrom Katz, *The Short Run and the Long Walk* (Santa Monica, Calif.: RAND, 1966), 4.

10. Andre Gunder Frank, "The Cold War and Me," *Bulletin of Concerned Asian Scholars* 29, no. 4 (1997): 79–84.

11. Rowen (December 11, 1968), 5–6. Some observers might say that instability is not so much a threat to democracy as a manifestation of it.

12. "Intelligence Panel Finds FBI and Other Agencies Violated Citizens' Rights," *New York Times*, April 29, 1976, 1; Guy Pauker, *Black Nationalism and Prospects for Violence in the Ghetto* (Santa Monica, Calif.: RAND, 1969).

13. Jardini (1996). In line with these goals, DoD created job programs for black men returning from Vietnam: Robert McNamara, "Social Inequities: Defense Department Programs Address, November 7, 1967," *Vital Speeches* 34 (December 1, 1967): 98–103.

14. H. L. Nieburg, "The Tech-fix and the City," in Herny Schmandt and Warner Bloomberg, eds., *The Quality of Urban Life, Urban Affairs Annual Reviews*, no. 3 (Beverly Hills, Calif.: Sage, 1969), 213.

15. Herbert Dordick, Leonard Chesler, Sidney Firstman, and Rudy Bretz, *Telecommunications in Urban Development* (Santa Monica, Calif.: RAND, 1969), v.

16. William Knox, "Problems of Communication in Large Cities," in Ithiel de Sola Pool, ed., *Talking Back: Citizen Feedback and Cable Technology* (Cambridge: MIT Press, 1973), 103.

17. Louis Wirth, *The Ghetto* (Chicago: University of Chicago Press, 1928); Michael Harrington, *The Other America: Poverty in the United States* (Baltimore: Penguin, 1963); Wellman (1974), Edmund Midura, ed., *Blacks and Whites: The Urban Communication Crisis: Why Aren't We Getting Through?* (Washington, D.C.: Acropolis Books, 1971); James Ettema and F. G. Kline, "Deficits, Differences, and Ceilings: Contingent Conditions for Understanding the Knowledge Gap," *Communication Research* 4 (1977): 179–202; Bradley Greenberg, *Use of the Mass Media by the Urban Poor* (New York: Praeger, 1970); Kerner Commission [National Advisory Commission on Civil Disorders], *Report* (New York: Dutton, 1968); Benjamin Bagdikian, "The Role of the News Media in the Urban Crisis," in Anthony H. Pascal, ed., *Contributions to the Analysis of Urban Problems* (Santa Monica, Calif.: RAND, 1968).

18. Dordick et al. (1969); Michael Shamberg, *Guerilla Television* (New York: Holt, Rinehart & Winston, 1971); Monroe Price and John Wicklein, *Cable Television: A Guide for Citizen Action* (Philadelphia: Pilgrim Press, 1972); National Academy of Engineering, *Communications Technology for Urban Improvement* (Washington, D.C.: National Academies, 1971); Charles Tate, ed., *Cable Television in the Cities: Community Control, Public Access, and Minority Ownership* (Washington, D.C.: Urban Institute, 1971); Knox (1973).

19. Edmund Burke, "Citizen Participation Strategies," *JAIP* 34 (September 1968): 287–94; Sherry Arnstein, "A Ladder of Citizen Participation," *JAIP* 35 (July 1969): 216–24; Neil Gilbert and Joseph Eaton, "Who Speaks for the Poor?" *JAIP* 36 (November 1970): 411–16; and Hans Spiegel, ed., *Citizen Participation in Urban Development* (Washington, D.C.: Center for Community Affairs, National Institute for Applied Behavioral Science, 1968).

20. Memo to all institute personnel from Joel Edelman, March 13, 1970, "Institute Seminar on Broadening Our Information Sources," no box, folder NYC Rand Institute 1970, RAND.

21. Seymour Schwartz, "Computers, Values, and Air Pollution Decisions," in *Proceedings of the Association for Computing Machinery Annual Conference on the Application of Computers to the Problems of Urban Society* (New York: ACM, 1971): 59–72.

22. Robert Yin, *Cable on the Public's Mind* (Santa Monica, Calif.: RAND, 1972), 4.

23. Ralph Engelman, "The Origins of Public Access Cable Television, 1966–1972," *Journalism Monographs* 123 (1990), 22.

24. Nicholas Johnson, *How to Talk Back to Your Television Set* (Boston: Little, Brown, 1970).

25. Harold Barnett and Edward Greenberg, *A Proposal for Wired City Television* (Santa Monica, Calif.: RAND, 1967) coined the phrase *wired city* (a term used frequently in this

era to mean a city with cable infrastructure throughout) as part of a contract project for RAND. Journalist Ralph Lee Smith—in "The Wired Nation," *Nation* (May 18, 1970): 582–606—coined the popular term *wired nation*.

26. Yin, *Cable on the Public's Mind* (1972), 4; Robert Yin and Grace Polk, *WNYC TV: Implications for New Cable Television Systems* (Santa Monica, Calif.: RAND, 1972); Rudy Bretz, *Television and Ghetto Education: The Chicago Schools Approach* (Santa Monica, Calif.: RAND, 1969); John Macy Jr., "Community Uses of Public Television" *City* 5, no. 2 (March/April 1971): 23–25; and Boston College and WGBH, *City in Crisis: An Evaluation of the Use of Television in Public Affairs* (Chestnut Hill, Mass.: Seminar Research Bureau, Boston College, 1954).

27. Jim Castelli, "Cable TV—a 'Common Carrier' or Not?" *America* (November 13, 1971): 397–400; *City Problems: The Annual Proceedings of the United States Conference of Mayors* (Washington, D.C.: United States Conference of Mayors, 1966), 169; and Mary Alice Mayer Phillips, *CATV: A History of Community Antenna Television* (Evanston, Ill.: Northwestern University Press, 1972).

28. Equating technology with democracy, increased citizen participation, economic development, and an antidote to the alienation of urbanization, predates the cold war. One can find claims about the more equal future to be delivered to Americans via electricity, telephone, or radio. Yet it was during the cold war that the idea to use communication technology as part of programmatic efforts to direct social and economic development first received simultaneous support from the U.S. military and intelligence communities and also advocates for social justice.

29. "Power to the People," in Leonard Zacks and Craig Harris, *The Instant Referendum—a CATV Based Direct Democratic Legislative Structure for Local Government* (Santa Monica, Calif.: RAND, 1971), 21.

30. Samuel Yette, *The Choice: The Issue of Black Survival in America* (New York: Putnam, 1971).

31. Norman Thomas and Harold Wolman, "Policy Formulation in the Institutionalized Presidency," in Thomas Cronin and Sanford Greenberg, eds., *The Presidential Advisory System* (New York: Harper & Row, 1969).

32. Cited in Melvin Webber, "The Roles of Intelligence Systems in Urban Systems Planning," *JAIP* 31, no. 4 (1965): 296 n. 3.

33. President's Task Force on Communications Policy, *Staff Paper Six* (Washington, D.C.: President's Task Force on Communications Policy, 1968), appendix B, *Telecommunications in Urban Development*, 141–42.

34. National Academy of Engineering (1971), vi.

35. Sloan Commission on Cable Communications, *On the Cable: The Television of Abundance* (New York: McGraw-Hill, 1971), 1.

36. Sloan Commission on Cable Communications (1971), 113–14.

37. Susan Douglas, "The Navy Adopts the Radio, 1899–1919," in Merritt Roe Smith, ed., *Military Enterprise and Technological Change* (Cambridge: MIT Press, 1985).

38. Leland Johnson, *Communications Satellites and Less-developed Countries* (Santa Monica, Calif.: RAND, 1967); Leland Johnson, *Some Implications of New Communications Technologies for National Security in the 1970s* (Santa Monica, Calif.: RAND, 1967); Guy Pauker, *Sources of Turbulence in the New Nations* (Santa Monica, Calif.: RAND, 1962); Guy Pauker, *Notes on Non-military Measures in Control of Insurgency* (Santa Monica, Calif.: RAND, 1962); Edward Hearle, *The Use of Computers in Government Operations of De-*

veloping Countries (Washington, D.C.: NTIS, 1970); L. D. Attaway, Herbert Dordick, and RAND Corp., *The Role of Technology in International Development* (Santa Monica, Calif.: RAND, 1966); Herbert Dordick, *Anonymous Communications: Some Thoughts on Pacification and Protection in Insurgency Areas and the Use of Personal Communication Devices* (Santa Monica, Calif.: RAND, 1964).

39. Nathan Leites and Charles Wolf, *Rebellion and Authority: An Analytical Essay on Insurgent Conflicts* (Chicago: Markham, 1970).

40. Suzanne Keller, *The American Lower Class Family* (Albany: New York State Division for Youth, 1968); Guy Pauker, *Black Nationalism and Prospects for Violence in the Ghetto* (Santa Monica, Calif.: RAND, 1969).

41. Memo to H. S. Rowen from P. L. Szanton: "Briefing for the President's Science Advisory Council, March 3, 1969," no box, folder NYC Project—1968, RAND.

42. Itiel de Sola Pool, "The Public and the Polity" (1967), reprinted in Lloyd Etheredge, ed., *Politics in Wired Nations: Selected Writings of Ithiel de Sola Pool* (New Brunswick, N.J.: Transaction, 1998), 266–67.

43. Robert Yin and Douglas Yates, *Street-level Governments: Addressing Decentralization and Urban Services* (Santa Monica, Calif.: RAND, 1974); Peter Steinberger, *Ideology and the Urban Crisis* (Albany: State University of New York, 1985).

44. Ithiel de Sola Pool and Herbert Alexander, "Politics in a Wired Nation," in Pool, *Talking Back*, 83; Colligan, in MITRE Corporation, *Symposium on Urban Cable Television* (Washington, D.C.: MITRE Corporation Washington Operations, 1973), vol. 3, 221–23; Robert Yin, *Some Remarks on Evaluating Administrative Decentralization* (Santa Monica, Calif.: RAND, 1972); Robert Yin, Brigitte Kenney, and Karen Possner, *Neighborhood Communications Centers* (Santa Monica, Calif.: RAND, 1974); Robert Yin and William Lucas, *Decentralization and Alienation* (Santa Monica, Calif.: RAND, 1973).

45. Etheredge (1998), 293.

46. Etheredge (1998), 299.

47. Daniel Patrick Moynihan, *Coping: Essays on the Practice of Government* (New York: Random House, 1973), 270.

48. Charles Silberman, *The Crisis in Black and White* (New York: Random House, 1964); Frances Fox Piven and Richard Cloward, *Regulating the Poor; the Functions of Public Welfare* (New York: Vintage, 1972).

49. NASA funded a number of research projects on cable, including papers appearing in Pool's collection *Talking Back*.

50. Laurence Lynn, personal conversation with author (Chicago, May 18, 2001). Like their occasional identification with the poor, this was at some level a false consciousness.

51. *City Problems* (1966), 48.

52. Robert Yin, *Cable Television: Citizen Participation in Planning* (Santa Monica, Calif.: RAND, 1973), 10); Peter Marris and Martin Rein, *Dilemmas of Social Reform* (New York: Atherton 1967); Fred Powledge, *Model City* (New York: Simon & Schuster, 1970); and Haar (1975).

53. Paul Ryan, *The Cybernetics of the Sacred* (New York: Anchor Doubleday, 1975); Price and Wicklein (1972); Fred Powledge, *An ACLU Guide to Cable Television* (New York: ACLU, 1972); Ted Ledbetter and Gilbert Mendelson, *The Wired City: A Handbook on Cable Television for Local Officials* (Washington, D.C.: Urban Communications Group, 1972).

54. Peter Goldmark, "Tomorrow We Will Communicate to Our Jobs," *Futurist* (April 1972): 55–58; Peter Goldmark, "Communication and the Community," *Scientific American* (September 1972): 143–50.

55. Kenneth Stetten and John Volk, *A Study of the Technical and Economic Considerations Attendant on the Home Delivery of Instruction and Other Socially Related Services via Interactive Cable TV,* vol. 1 (Washington, D.C.: MITRE, May 1973), 47.

56. Leland Johnson confirms that there was collegial interaction between these constituencies. According to Johnson, who spent most of his career at RAND, the key difference was that people at such institutions as the Alternate Media Center "thought of us as ivory tower and intellectual": telephone conversation with author, September 13, 2001.

57. Dordick et al. (1969), 141–42; Benjamin Bagdikian and Kathleen Archibald, *Televised Ombudsman* (Santa Monica, Calif.: RAND, 1968); MITRE (1973), vol. 3, 248.

58. Deirdre Boyle, *Subject to Change: Guerilla Television Revisited* (New York: Oxford University Press, 1997).

59. MITRE (1973), vol. 1, 36.

60. Daniel Patrick Moynihan, *Maximum Feasible Misunderstanding: Community Action in the War on Poverty* (New York: Free Press, 1969), 178; Lasch (1969); Lindsay Papers, subject files 1966–73, box 23, folder 394, Press (National) 1966–73, where Lindsay kept a list of national press coverage on a variety of topics including "JVL's Shirtsleeve Ghetto Diplomacy," MANY. This identification between elites and the poor is hilariously satirized in Tom Wolfe, *Radical Chic & Mau-Mauing the Flak Catchers* (New York: Farrar, Straus & Giroux, 1970).

61. Martin Greenberger, ed., *Computers, Communications, and the Public Interest* (Baltimore: Johns Hopkins Press, 1971).

62. "For the record," from R. Yin: "Visit to Yale's Urban Studies Program," 3 December 1970, 2, no box, folder NYC Rand Institute 1970, RAND.

63. Memo to Peter Szanton from Roberta Goldstein, "Meeting with an Antimilitary Militant," July 16, 1969, no box, folder NYC Rand Institute 1969, RAND.

SEVEN: Wired Cities

1. Herman Kahn, in MITRE Corporation, *Symposium on Urban Cable Television* (Washington, D.C.: MITRE Washington Operations, 1973), 1:48.

2. Kahn, MITRE (1973), 1:45. There is some irony here: this was the same sort of critique that earlier had been leveled against nuclear strategists, including Kahn—that they were trying, mathematically, to rationalize an irrational situation.

3. Bernard Frieden and Lynne Sagalyn, *Downtown, Inc.: How America Rebuilds Cities* (Cambridge: MIT Press, 1989), 87.

4. Daniel Patrick Moynihan, *Coping: Essays on the Practice of Government* (New York: Random House, 1973), 273–74.

5. RAND, Twentieth Year Conferences, NYC special session, March 11, 1969, the University Club, RAND, box NYC RAND Institute, folder Twentieth Year Conferences, 28, RAND.

6. Anthony Mancini, "Little City Halls Called Big on Cooling Ghettos," *New York Post,* April 17, 1968, n.p., Lindsay Papers, subject files 1966–73, box 62, folder 1171, Lit-

tle City Halls 1968; Betty Flynn, "To 'Cool' Troubled Areas Mayor Lindsay Cozies up to Militant Youth Leaders," *Fargo Forum* June 14, n.p., Lindsay Papers, subject files 1966–73, box 23, folder 394, Press (National) 1966–73, MANY.

7. Cover letter accompanying Mayor's Advisory Task Force on CATV and Telecommunications, *A Report on Cable Television and Cable Telecommunications in New York City* (New York: Mayor's Advisory Task Force, September 1968).

8. Sloan Commission (1972), appendix C, n.p.

9. Ralph Engelman, "The Origins of Public Access Cable Television, 1966–1972," *Journalism Monographs* 123 (1990).

10. David Othmer, *The Wired Island* (New York: Fund for the City of New York, September 1973), VF NAC 8585g27 NewY F 1973.

11. Price and Wicklein (1972), 102; Othmer (1973); Engelman (1990); and Alan Wurtzel, "Public Access Cable TV: Programming," *Journal of Communication* 25, no. 3 (1975).

12. Engelman (1990), 34.

13. Othmer (1973), 4; The *Loretto vs. Teleprompter* court case (1972) raised the question as to whether cable is a "public space."

14. Othmer (1973), 32.

15. Carol Anschien, *Public Access Celebration: Report on Public Access in New York* (New York: Glad Day Press, 1973), 15.

16. Anschien (1973).

17. Edward Blum, *The Community Information Utility and Municipal Services* (Santa Monica, Calif.: RAND, 1972).

18. Othmer (1973), 8. The office is now known as the Department of Information Technology and Telecommunications.

19. Robert Yin, *Cable Television: Applications for Municipal Services* (Santa Monica, Calif.: RAND, 1973).

20. Letter to Mayor Lindsay from Herbert Dordick, May 24, 1973, Lindsay Papers, subject files 1966–73, Dordick, Herbert S., 1973, MANY.

21. Simpson (1994), Edward Berman, *The Ideology of Philanthropy: The Influence of the Carnegie, Ford, and Rockefeller Foundations on American Foreign Policy* (Albany: State University of New York Press, 1983).

22. "National League of Cities and US Conference of Mayors and the City of New York, Conference on Cable Television, February 5–6, 1973," Lindsay Papers, subject files 1966–73, box 28, folder 403, Franchises, Bureau of, 1973, MANY.

23. Herbert Dordick and Frank Young, cited in Kas Kalba, *City Meets the Cable: A Case Study in Technological Innovation and Community Decision-making* (Cambridge: Harvard University Program on Information Technologies and Public Policy, 1975), 159–60.

24. Peg Kay and Stanley Gerandasy, *Social Services and Cable TV* (Washington, D.C.: NSF, 1976).

25. LA City contracts 128876, 129053, 132349, respectively, MALA. Figures from Herbert Dordick, *Cable Communications in Los Angeles Master Plan* (Los Angeles: LA Department of Public Utilities and Transportation, 1976), and Herbert Dordick, Frederick Byrne, and Alan Talbot, *A Study of Local Government Uses of Cable Television and Integrated Communication Systems* (Los Angeles: University of Southern California Annenberg School of Communications, 1975), table 1.

26. *Summary Report of the Los Angeles Goals Council* (Los Angeles: Goals Council, November 1969), 72–73, VF NAC 6827 LosA L; Ann Howell, *City Planners and Planning in Los Angeles, 1781–1998* (Los Angeles: Department of City Planning, June 1998).

27. *Summary Report of the Los Angeles Goals Council* (November 1969), 72–73.

28. Leonard Chesler and Herbert Dordick, *Communications Goals for Los Angeles, a Working Paper for the Los Angeles Goals Program* (Santa Monica, Calif.: RAND, 1968), 7.

29. Eleanor McKinney, ed., "Pacifica's License Renewal Controversy," in *The Exacting Ear: The Story of Listener-sponsored Radio, and an Anthology of Programs from KPFA, KPFK, and WBAI* (New York: Pantheon, 1966).

30. Don Bushnell, *System Simulation: A New Technology for Education* (Santa Monica, Calif.: SDC, 1962); Don Bushnell and Hallock Hoffman, "The Watts Community Communications Bureau and Training Center: A Proposal for the Mafundi Institute," March 1970, unpublished; Richard Kletter, *Cable Television: Making Public Access Effective: Prepared for the National Science Foundation* (Santa Monica, Calif.: RAND, 1973); Watts Communication Bureau, *An Application to the City of Los Angeles for a Franchise to Install and Operate a Community Antenna Television System in the Los Angeles Basin* (Los Angeles: Watts Communication Bureau, 1972), LA council file 71-2558 (and supplements), MALA.

31. Herbert Dordick and Jack Lyle, *Access by Local Political Candidates to Cable Television* (Santa Monica, Calif.: RAND, 1971).

32. Leonard Zacks and Craig Harris, *The Instant Referendum—a CATV Based Direct Democratic Legislative Structure for Local Government* (Santa Monica, Calif.: RAND, 1971).

33. Robert Warren, ed., *The Wired City of Los Angeles: Papers and Discussions from a Seminar on Urban Cable Television* (Los Angeles: USC Center for Urban Affairs, 1972), iii.

34. Warren (1972), 4.

35. Warren (1972), 50.

36. Dordick (1976), 27–28.

37. Dordick, Byrne, and Talbot (1975), 28.

38. Remarks in Warren (1972), 17–19.

39. Herbert Dordik and Nicholas Valenzuela, *A Study of Mexican-American Information Sources and Media Usage in Boyle Heights (East Los Angeles)* (Los Angeles: University of Southern California Annenberg School of Communications, 1974); Dordick, Byrne, and Talbot (1975), n.p., n. 39.

40. Nonetheless, according to Leland Johnson (telephone conversation with author, September 13, 2001), relations among these groups were favorable and contacts were frequent.

41. Kalba (1975).

42. Johnson, telephone conversation, September 2001.

43. Harold Barnett and Edward Greenberg, *A Proposal for Wired City Television* (Santa Monica, Calif.: RAND, 1967).

44. Leland Johnson, Walter Baer, Rudy Bretz, Donald Camph, Ned Feldman, Rolla Park, and Robert Yin, *Cable Communications in the Dayton–Miami Valley: Basic Report* (Santa Monica, Calif.: RAND, 1972); Ann Kline, "RAND Report on a Regional vs. City CATV in Dayton Area Cues Conflicts; Push Interconnected 6-Cable System," *Variety* (February 16, 1972); Ned Feldman, *Cable Television: Opportunities and Problems in Local program Origination* (Santa Monica, Calif.: RAND, 1970); Robert Yin, "Cable TV and Public Interest Programs in Dayton, Ohio," unpublished, August 1971, RAND; Robert Yin,

"Television and the Dayton Resident: The Results of a Public Opinion Survey," unpublished, 1971, RAND.

45. RAND's later Spartanburg experiment was not a franchising project but a test of two-way services for adult education and parenting education.

46. Marion Hayes Hull, "Economic Potential for Minorities—Obstacles and Opportunities," in Mary Louise Hollowell, ed., *The Cable/Broadband Communications Book, 1977–1978: A Guide to Cable and New Communication Technologies* (Washington, D.C.: Communication Press, 1977), 80; Marion Hayes Hull, "Minorities," in Mary Louise Hollowell, ed., *Cable Handbook, 1975–1976: A Guide to Cable and New Communication Technologies* (Washington, D.C.: Publi-Cable, 1975).

47. Davis Dyer and Michael Aaron Dennis, *Architects of Information Advantage: The MITRE Corporation since 1958* (Montgomery, Ala.: Community Communications, 1998).

48. Peggy Karp, "MITRE Network Working Group: Request for Comments 321" (March 24, 1972): rfc/std/fyi/bcp; "Going Digital: MITRIX," *MITRE Matrix* 5, no. 6 (December 1972): 38–56; E. K. Smith, "Pilot Two-way CATV Systems," *IEEE Transactions on Communications* 23 (January 1975): 111–20; and D. G. Willard, "Mitrix: A Sophisticated Digital Cable Communications System," in *Proceedings of the National Telecommunications Conference* (New York: IEEE, November 1973), vol. 2, 38E/1–5.

49. Kenneth Stetten and John Volk, *A Study of the Technical and Economic Considerations Attendant on the Home Delivery of Instruction and Other Socially Related Services via Interactive Cable TV*, vol. 1 (Washington, D.C.: MITRE, May 1973), 1; William Mason, Frank Eldridge, John O'Neill, Carol Paquette, Sidney Polk, Friend Skinner, and Ralph Lee Smith, *Urban Cable Systems* (Washington, D.C.: MITRE, May 1972), ii–5.

50. Michael Herrero, *New Communities and Telecommunications* (Chapel Hill: Center for Urban and Regional Studies, University of North Carolina, 1973), VF NAC 8585 H 1973.

51. Stetten and Volk (May 1973), 1:41.

52. Stetten and Volk (May 1973), 1:37–38.

53. MITRE, *Urban Cable Systems* (MacLean, Va.: MITRE, November 1971), 55; *Testing the Applicabilities of Existing Telecommunications Technology in the Administration and Delivery of Social Services* (MacLean, Va.: MITRE, April 1973).

54. Stetten and Volk (May 1973), 1:47.

55. MITRE (November 1971), 6, 55.

56. In 1976, Warner Cable did develop the two-way QUBE system in Columbus, Ohio, but this was not a widespread use for the medium.

57. Martin Topper and Leigh Wilson, "Cable Television: Applied Anthropology in a New Town," *Human Organization* 35, no. 2 (1976): 135–46; Jerrold Oppenheim, "The Wonders of Rewiring America" *Progressive*, June 1972, 22–23.

58. Thomas Grubisich, "Reston Cable TV Still Struggling to Exist," *Washington Post*, March 21, 1977, C1.

59. Kay and Gerandasy (1976).

60. Dickson (1971).

61. William Gorham, "Notes of a Practioner," *The Public Interest* 8 (summer 1967): 4–8.

62. Specifically, Ted Ledbetter and Gilbert Mendelson, *The Wired City: A Handbook on Cable Television for Local Officials* (Washington, D.C.: Urban Communications Group, 1972).

63. Leland Johnson recalls frequent conversations between RAND and Urban Institute staff: telephone conversation, September, 2001.

64. Tate, in Ithiel de Sola Pool, ed., *Talking Back: Citizen Feedback and Cable Technology* (Cambridge: MIT Press, 1973), 54, 60; Charles Tate, "Cable TV: Implications for Black Communities," *Urban League News* (September 4, 1973).

65. MITRE (1973), 1:9.

66. MITRE (1973), 3:206.

67. Marvin Rimmerman, "Cable and the Baltimore City Priority: Director, Mayor's Office of Telecommunications," *TV Communications* (November 1974): 30–37.

68. Federal agencies (HUD, HEW, NSF) explored innovative services by supporting demonstrations on a case-by-case basis. Grant recipients (including Ithiel de Sola Pool, W. Bowman Cutter, and Herbert Dordick) and their projects are described in several issues of the *Urban Telecommunications Forum,* 1973 and 1974.

69. MITRE (1973), 4:53–55.

70. Deirdre Boyle, *Subject to Change: Guerilla Television Revisited* (New York: Oxford University Press, 1997); National Cable Television Association, *Over the Cable* (Washington, D.C.: National Cable Television Association, 1974).

71. Kas Kalba, *City Meets the Cable: A Case Study of Cincinnati's Decision Process* (Cambridge: Harvard University Program on Information Technologies and Public Policy, 1974); Robert Steiner, *Visions of Cablevision* (Cincinnati: Stephen H. Wilder Foundation, December 1972).

72. Kay and Gerandasy (1976), ii–36.

73. Kalba (1975), 224.

74. Kalba, cited in Dordick, Byrne, and Talbot, (1975), 6.

75. *FCC v. Midwest Video Corp,* 440 US 689. There has, however, been one widely successful use of two-way cable systems: for home security.

76. Thomas Hazlett, "Station Brakes: The Government's Campaign against Cable Television," *Reason Online* 26, no. 9 (1995): 40–47.

77. Steven Lohr, "Data Highway Ignoring the Poor, Study Charges," *New York Times,* May 24, 1994, A1.

78. Manuel Castells, *The Informational City* (Oxford: Blackwell, 1989); and Manuel Castells, *The Rise of the Network Society* (Oxford: Blackwell, 1996).

79. John J. O'Connor, "TV View: Fluff Clouding Cable's Future," *New York Times,* July 25, 1982, sec. 2, 21; Marcia A. Slacum, "Dial Dilemma: Would District Subscribers Actually Watch as Many as 152 Cable TV Channels?" *Washington Post,* April 15, 1984, C1.

Conclusion

1. Christopher Simpson, *Science of Coercion* (New York: Oxford University Press, 1994); Christopher Simpson, *Universities and Empire: Money and Politics in the Social Sciences during the Cold War* (New York: Free Press, 1998); Peter Galison and Bruce Hevly, eds., *Big Science: The Growth of Large-scale Research* (Stanford, Calif.: Stanford University Press, 1992); Daniel Kevles, *The Physicists: The History of a Scientific Community in Modern America* (New York: Knopf, 1977); Richard Rhodes, *The Making of the Atomic Bomb* (New York: Simon & Schuster, 1986).

2. William Fulbright, "Militarism and American Democracy," Owens-Corning Lecture to Denison University, April 8, 1969, reprinted in L. Freedman, ed., *Issues of the Seventies* (Belmont, Calif.: Wadsworth, 1970), 362.

3. Stokely Carmichael and Charles Hamilton, *Black Power: The Politics of Liberation in America* (New York: Random House, 1967).

4. This phenomenon can be found earlier than the 1940s, but it picked up speed in World War I. See Simpson (1994) for a discussion of how social scientists, especially communication researchers, participated in World War I.

5. "Other defense research installations" refers to the defense and space agencies' in-house research centers, such as the Human Resources Research Institute at the Maxwell Air Force Base.

6. D. M. Simpson, "Virtual Reality and Urban Simulation in Planning: A Literature Review and Topical Bibliography," *Journal of Planning Literature* 15, no. 3 (2001), 359–72; Paul Waddell, "UrbanSim: Modeling Urban Development for Land Use, Transportation, and Environmental Planning," *Journal of the American Planning Association* 68, no. 3 (2002); Paul Starr, "Seductions of Sim: Policy as a Simulation Game," *American Prospect* 5, no. 17 (1994): 19–29; "Complex Systems and Policy Analysis: New Tools for a New Millennium," RAND Corporation Conference, September 27 and 28, 2000: rand.org/scitech/stpi/Complexity/; Michael E. Wiseman, *A Management Information Model for New-Style Public Assistance* (Washington, D.C.: Urban Institute, 1999); Jonathan Rauch, "Seeing Around Corners," *Atlantic Monthly* 289, no. 4 (2002): 35–48; Perry Hardin, "Remote Sensing/GIS Integration to Identify Potential Low-Income Housing Sites" *Cities* 17, no. 2 (2000): 97–109; Sarah Elwood and Helga Leitner, "GIS and Community-based Planning: Exploring the Diversity of Neighborhood Perspectives and Needs," *Cartography and Geographic Information Systems* 25, no. 2 (1998): 77–88; Steve LeSueur, "Nimble Upstarts Clamor for Role in State, Local E-services," *Washington Technology* (November 22, 1999), 1; Greg Schneider, "Lockheed Unit Wins Big Contract: D.C.-Based IMS Gets Orange County Award," *Washington Post,* November 17, 2000, E05; Jennifer Light, "Rethinking the Digital Divide," *Harvard Educational Review* 71, no. 4 (2001): 709–33.

7. Indeed, in 1994, for example, the *Journal of the American Planning Association* 60, no. 1, revisited Douglass Lee's 1973 "requiem" for urban models. While the contributors to this theme issue repeatedly noted the historical longevity of urban modeling, none mentioned the military origins of these tools.

8. With the possible exception of policing and home-security protection.

9. Anthony Flint, "September 11 Pushes Firms to the Suburbs," *Boston Globe,* August 18, 2002, A1, A16; Todd Milbourn, "Minnetonka Mayor Seeks More Money for Cities to Offset New Security Costs," *Minneapolis Star Tribune,* January 16, 2002, 6A; James Dao, "Internal Security is Attracting a Crowd of Arms Contractors," *New York Times,* March 20, 2002, C1, C11; George W. Bush, executive order establishing the Office of Homeland Security (October 8, 2001); David Broder, "Mayors Seek Clear Security Plan," *Washington Post,* June 18, 2002, A10.

Note on Sources

PRIMARY SOURCES

Unpublished and Limited-Circulation Sources

The unpublished and limited-circulation sources cited throughout this book may be found in four collections:

Harvard University's Graduate School of Design Library Vertical Files Collection contains numerous materials on topics discussed throughout the book. These include speeches by Tracy Augur, mimeographed government reports of the federal Industrial Dispersion Program, bulletins from the Pittsburgh Community Renewal Program and the Los Angeles Community Analysis Bureau, advertising brochures from aerial survey companies, bibliographic compilations of research reports on planning for civil defense, research papers from staff at the Harvard Laboratory of Computer Graphics and Spatial Analysis, and miscellaneous newspaper clippings related to the history of American city planning. Each source is individually cataloged on Harvard's library computer system (holliscatalog.harvard.edu).

RAND Corporation headquarters in Santa Monica opened its archives in the late-1990s, making available to researchers records of the New York City RAND Institute, including correspondence between RAND and city staff, draft versions of reports, and newspaper clippings documenting RAND's work for that city. Also worth special mention is the internal RAND library computer system, accessed on site, which catalogs many reports available only internally. Searching this database yields a far more detailed picture of ongoing research at the think tank than do searches of any other bibliographic database, including RAND's own publicly accessible publications website www.rand.org/publications. (*Note:* All RAND materials cited in this project were unclassified or declassified documents; RAND maintains a separate library and computer system for classified materials.)

The city archives in New York and Los Angeles offer ample information on activities behind the scenes as municipal governments tried to adopt new technologies. At the Municipal Archives of the City of New York, John V. Lindsay's papers are especially well cataloged; available for consultation on microfilm, the records of the Lindsay administration's dealings with RAND and NASA are easy to locate. Of special interest at the Municipal Records Center of the City of Los Angeles are files with information on all city contracts, which may be used to document relationships between city agencies and aerial survey companies, applications from cable operators to wire city neighborhoods, and contracts with independent consultants hired to assist city agencies implementing new technologies. Draft and final reports from the Los Angeles Community Analysis Bureau are also held there.

Journals and Periodicals

Academic journals and trade publications of several professional organizations provide insights into how military and urban planners and managers saw opportunities, and occasional threats, in their emerging collaborations. Relationships between atomic scientists, defense experts, and city administrators during the late-1940s and early-1950s are especially vivid in issues of the *Journal of the American Institute of Planners* (later, the *Journal of the American Planning Association*) and the *Bulletin of the Atomic Scientists* for the late-1940s and early-1950s.

For the 1960s and 1970s, the *Journal of the American Institute of Planners* and the *Public Administration Review* document the continued appeal of defense and aerospace techniques and technologies in American city planning and management. Also noteworthy in this period is the trade weekly *Aviation Week and Space Technology;* its pages are filled with industry leaders' confessed anxieties about the future of the industry, as well as their hopes for new markets in urban problem solving.

Conference Proceedings

Researchers tracking the circulation of ideas from defense and aerospace to urban government will find ample documentation in the published proceedings of conferences bringing together individual and institutional representatives of both communities. Of particular interest are proceedings of annual meetings of the Association of Computing Machinery Conference on the Application of Computers to the Problems of Urban Society; the Urban and Regional Information Systems Association; and the American Institute of Aeronautics and Astronautics/Public Technology Inc. conferences on Urban Technology.

One-time meetings with proceedings of note include the MITRE Corporation's *Symposium on Urban Cable Television: October 18th, 19th, and 20th, 1972, Sponsored by the Mitre Corporation* (Washington, D.C.: MITRE Corporation Washington Operations, 1973); Robert Warren, ed., *The Wired City of Los Angeles: Papers and Discussions from a Seminar on Urban Cable Television* (Los Angeles: USC Center for Urban Affairs, 1972; the U.S. National Aeronautics and Space Administration's *Conference on Space, Science, and Urban Life: Proceedings of a Conference Held at Oakland, California, March 28-30, 1963, Supported by the Ford Foundation and the National Aeronautics and Space Administration in Cooperation with the University of California and the City of Oakland* (Washington, D.C.: GPO, 1963); and the American Academy of Political and Social Science's *Governing Urban Society: New Scientific Approaches* (Philadelphia: AAPSS, 1967).

SECONDARY MATERIALS

The search for secondary materials in preparing this book provided a case study of how academic disciplines analyzing the same historical era can abstract divergent sets of themes. Looking back on the U.S. experience during the cold war, historians of science and technology, as well as military historians, see a period of innovation in techniques and technologies, a theme generally absent from urban histories of the period. Several studies of the development and uses of the innovations described throughout this book stand out for special mention.

A clear introduction to the history and uses of war gaming and computer simulations is John Raser's *Simulation and Society: an Exploration of Scientific Gaming* (Boston: Allyn & Bacon, 1969), which clarifies distinctions among information theory, cybernetics, operations research, and other related terms that are often used interchangeably. Paul Edwards, *The Closed World* (Cambridge: MIT Press, 1996) documents the military histories of cybernetics and computer simulations, exploring the implications when war planners' vision of the world as a closed system diffused beyond the boundaries of military and political analysis. Agatha Hughes and Thomas Hughes, eds., *Systems, Experts, and Computers* (Cambridge: MIT Press, 2000) presents an overview of the expanding appeal of systems analysis to multiple professional communities during the cold war. Thomas Hughes, *Rescuing Prometheus* (New York: Pantheon Books, 1998), an account of several large-scale projects of the military-industrial complex, touches upon several of the urban experiments discussed in this book.

The history of military reconnaissance from air and space is a robust area of research, despite much of the historical record remaining classified. Technological developments for surveillance and reconnaissance from space are especially thoroughly documented in Jeffrey Richelson, *The U.S. Intelligence Community* (Boulder, Colo.: Westview Press, 1999) and William Burrows, *Deep Black* (New York: Random House, 1986). Pamela Mack's *Viewing the Earth* (Cambridge: MIT Press, 1990) is the classic history of Landsat. Mack, while focused on the civilian technology, devotes careful attention to the occasionally tense relationships between military and civilian space programs. Also of note is John Cloud's research on the recently declassified CORONA satellite and its impact on civilian mapping efforts at the USGS; see especially John Cloud, "Imaging the World in a Barrel: CORONA and the Clandestine Convergence of the Earth Sciences," *Social Studies of Science* 31, no. 2 (2001): 231–51; and John Cloud, "Re-viewing the Earth: Remote Sensing and Cold War Clandestine Knowledge Production," *Quest: The History of Space Flight Quarterly* 8, no. 3 (2000): 5–15.

Relationships between U.S. military and intelligence priorities and trends in scientific and social scientific research are the subject of several good studies. Stuart Leslie's *The Cold War and American Science: The Military-Industrial-Academic Complex at MIT* (New York: Columbia University Press, 1993) is one of the best sources on changes in university scientific research. Christopher Simpson's *Science of Coercion* (Oxford, England: Oxford University Press, 1994) is the premier account of linkages between military planners' concerns about national security and the history of communication research; Simpson focuses on psychological warfare studies, but he also features significant discussions of development theory. Michael Latham's *Modernization as Ideology: American Social Science and "Nation Building" in the Kennedy Era* (Chapel Hill: University of North Carolina Press, 2000), which focuses on modernization and development during the Kennedy administration, is especially strong in its documentation of connections between American social science theories and U.S. foreign policy applications. Paul Dickson's *Think Tanks* (New York: Atheneum, 1971) details the transition taking place in American think tanks

from exclusive or near-exclusive military contracting toward domestic research during the Johnson administration and the ways in which a variety of
contract defense research organizations turned their attention to civil systems
research. However, while Dickson suggests that at some institutions such as
SDC there was some animosity between the urban studies types and the military strategists, the evidence discussed here finds that many of the individuals
staffing the urban projects at SDC and other think tanks came to urban research with backgrounds in military contract research.

To date, the history of cable has been told, much like the history of other
communication technologies, as a story in which excitement about cable as
an instrument of democracy was eclipsed when a few powerful corporate interests grabbed hold of that technology. In these accounts, think tank participation is interpreted in terms of this larger narrative, rather than linked to
think tanks' other ongoing activities. Especially rich sources for documenting
the history of debates about cable television during the FCC freeze are Ralph
Engelman, "The Origins of Public Access Cable Television, 1966-1972," *Journalism Monographs* 123 (1990); Thomas Streeter, "Blue Skies and Strange Bedfellows: The Discourse of Cable Television," in Lynn Spigel and Michael Cur-
tin, eds., *The Revolution Wasn't Televised: Sixties Television and Social Conflict*
(New York: Routledge, 1997); and Deirdre Boyle, *Subject to Change: Guerilla
Television Revisited* (New York: Oxford University Press, 1997).

Within urban studies and allied disciplines, one major narrative about
militarization in the United States has developed since the 1970s—a line of argument focused on the fear of crime and the turn toward "security" in the nation's housing stock, from gated communities to private security to forms of
private residential government. In this research, the language of war is used
more figuratively than literally. The few sources documenting historical relationships between the nation's security needs and its cities focus primarily on
physical planning rather than on the postwar emergence of a new type of urban expert whose interests included social planning. Such sources deal with
topics including the architecture of armories and military bases and the role of
the siting of military industry in regional economic development and decline.
Examples are Robert Fogelson, *America's Armories* (Cambridge: Harvard University Press, 1989); Ann Markusen, Peter Hall, Scott Campbell, and Sabina
Dietrich, *The Rise of the Gunbelt* (New York: Oxford University Press, 1991); Michael Dudley, "Sprawl as Strategy: City Planners Face the Bomb," *Journal of
Planning Education and Research* 21 (2001): 52-63; and Donald Albrecht, ed.,

World War II and the American Dream: How Wartime Building Changed a Nation (Cambridge: MIT Press, 1995). For an appreciation of how California's urban history was inextricably linked to the development of the military-industrial complex, see Roger Lotchin, *Fortress California, 1910–1961* (New York: Oxford University Press, 1992). Lotchin uses the phrase *the metropolitan-military complex* to describe how city leaders used military and defense concerns to bolster the growth of their regions.

Most histories of the planning profession and collections of "essential readings" in the history and theory of city planning do not give their due to the defense intellectuals. Two exceptions include a reader of short essays in planning theory compiled by M. C. Branch, ed., *Urban Planning Theory* (Stroudsburg, Pa.: Dowden, Hutchinson & Ross, 1975), and John Friedmann's overview of planning, *Planning in the Public Domain* (Princeton, N.J.: Princeton University Press, 1987).

Finally, while their focus is neither the history of defense and aerospace technology nor the history of American cities, two historians have situated twentieth-century American poverty programs and civil-rights reforms in a larger context that includes the cold war. Mary Dudziak's *Cold War Civil Rights* (Princeton, N.J.: Princeton University Press, 2000) links the history of American civil-rights reforms (1940s to the 1960s) to political events on the international stage, emphasizing how an extension of limited civil rights to America's minority populations was considered to be in the strategic interest of the United States. Alice O'Connor's *Poverty Knowledge* (Princeton, N.J.: Princeton University Press, 2001) documents the history of poverty research and its relationship to social policy; her account describes the movement toward domestic research at RAND as part of a larger "poverty research industry" with a lasting legacy for poverty research and policy to the present day.

Index